Praise for
The American Plate

"*The American Plate* is an engagingly readable history of American food. It takes us from pemmican to microwave popcorn, from lunch pails to oysters Rockefeller. It's one of those agreeable books that works just as well if you dive in at random or read every bite in sequence; either way, it is full of treats. Libby O'Connell imparts a great deal of information about changing American foodways with humor and pithiness. She considers such questions as how beaver tail and eel—once popular American delicacies—didn't pass the test of time, while maple syrup and blueberries are still enjoyed, just as they were by American Indians. Oh, and read on to find out the secret ingredient in Brunswick stew."

—Bee Wilson, author of *Consider the Fork: A History of How We Cook and Eat*

"Like many miniencyclopedias, this one is studded with often intriguing facts…O'Connell is a perky companion for this buffet of historical snacks."

—*Kirkus*

"*The American Plate* shows that food is interesting not just because it tastes good, but because it reflects the people who grow, cook, and eat it. Here are the compelling, colorful backstories of national bites as familiar as microwave popcorn and as obscure as pemmican, as enduring as southern fried chicken and as of-the-moment as seaweed and acai—a buffet of information that revels in the incredible diversity of American cuisine."

—Michael Stern, coauthor of the bestselling Roadfood series

"Every page tells a surprising story about American culinary history. O'Connell's easy-to-digest prose and modest portions are sprinkled with fascinating anecdotes that make these stories compulsively readable. This book is a wonderful delight."

—Andrew F. Smith, editor in chief, *The Oxford Encyclopedia of Food and Drink in America*

"From maize and barbecue to granola and salsa, Libby O'Connell illuminates the history of America's palate with fun and engaging tales of how the nation became what it eats. Using anthropology, history, folklore, and economics as her ingredients, she cooks up an intellectually delicious buffet sure to please a wide readership. Thoroughly researched, and based upon serious scholarship, this is a wonderful, informative, and entertaining book."

—Richard Kurin, author of *The Smithsonian's History of America in 101 Objects*

the american plate

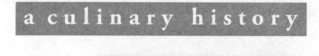

a culinary history

in 100 bites

LIBBY H. O'CONNELL

Published by Sourcebooks, Inc.
P.O. Box 4410, Naperville, Illinois 60567-4410
(630) 961-3900
Fax: (630) 961-2168
www.sourcebooks.com

Library of Congress Cataloging-in-Publication Data

O'Connell, Libby H.
  The American plate : a culinary history in 100 bites / Libby H. O'Connell.
    pages cm
  Includes bibliographical references and index.
  (hard cover : alk. paper) 1. Food—United States—History. 2. Cooking, American—History. I. Title.
  TX360.U6O35 2014
  641.5973—dc23
                            2014020381

Printed and bound in the United States of America.
WOZ 10 9 8 7 6 5 4 3 2

To Matt

CONTENTS

CHAPTER 3: BREAKFAST WITH RUM AND TEA

From Colonies to Country

55

CHAPTER 4: ROAST TURTLES AND HANGTOWN FRY

The Rise of a New Nation

81

CHAPTER 8: SCRAMBLED EGGS, HERSHEY BARS, AND PEACH COBBLER

The Great Depression and World War II

205

CHAPTER 9: COCA-COLA, ICEBERG LETTUCE, AND FAST FOOD

The Postwar, Cold War Era

231

CHAPTER 10: MICROWAVE POPCORN, MESCLUN GREENS, AND SALSA

1969–2000

259

EPILOGUE: A FEW EXTRA BITES

American Food Today

289

ACKNOWLEDGMENTS

I have many people to thank for their help with this book and my professional journey as a public historian. My professors and friends at the University of Virginia, particularly my adviser, Steve Innes, and Bruce Ragsdale, influenced my thinking as a historian so many years ago. The scholars and staff at organizations such as Plimoth Plantation, Mount Vernon, the Smithsonian, and Monticello have shared their expertise and enthusiasm for American history in its rich diversity. Thank you all for your intellectual and personal generosity. To food historian Andrew F. Smith and his many books, I owe a debt of gratitude for helping me navigate my way through the trees in order to find the forest. Thanks to David Wilk at Booktrix and Stephanie Bowen at Sourcebooks for believing in this project. The authors,

historians, and cooks whose work has guided me deserve my appreciative recognition. But please take note: all errors here are my own.

The smart, talented people at History/A+E Networks have had a profound impact on my life. I want to thank Abbe Raven, whose extraordinary leadership and business acumen combines with a commitment to corporate citizenship and friendship. Nancy Dubuc, a media visionary who brings excitement to the office every day, influenced me—and this book—by always encouraging me to think big. I admire Michael Feeney probably more than he realizes, for his perceptive take on life, his ability to work with anyone, and his unflappable good humor. To Charlie Maday, Whitney Goit, and Nick Davatzes for taking a chance on hiring an academic, and to Rob Sharenow and Pete Gaffney for writing their books while working at A&E, enabling me to see what's possible. Many thanks also to Mead, Kirsten, Lissette, Alannah, the Susans, Scott, Paul, John, David, Joan, Didi, Joy, Kat, Lynn, Vicky, Lindsay, and so many others. Fellow historian Kim Gilmore served as a thoughtful, expert adviser and helpful friend throughout this process. I'll always be indebted to you, Doctor G.

My dear friends in Brooklyn, Cold Spring Harbor, Desbarats, Singapore, and Washington, DC, have provided unflagging encouragement—life would be dreary indeed without our brisk walks, long talks, and wonderful dinners. Thank you.

My large extended family brings a lot of joy into my life, but I particularly want to thank my siblings, Terry Haight, Ginger Stevenson, and Kitty Petty, who have played larger roles than they realize in shaping this book. Vivid childhood memories of mundane and extraordinary meals and escapades with my brother and sisters arose while I wrote these chapters, although only a few made it into the final copy.

My mother, Debby Haight, did not inspire me with her cooking, but she did inspire me with her constant love of learning. And Jack Haight, my father, told the best history stories a child could imagine. Together they had a profound influence on my career as a historian and on my lifelong enthusiasm for sharing ideas.

Some of the most enjoyable moments in my life have centered around the table—breakfast, lunch, tea, and dinner—with my children, Charlie and

Lucy O'Connell. I'm lucky to be related to two remarkable adventurers, who are also interesting, kind, and a whole lot of fun. I love you very much.

And finally, I want to acknowledge my husband, Matt O'Connell, whose love and encouragement never wavered during the writing of this book. Making crème caramel for you on that cold day in November, 1975, turned out to be the best decision of my life.

✣ INTRODUCTION ✣

Imagine, for a minute, traveling to a foreign country and exploring that nation's culture. How might you hope to really understand the people—their traditions, their customs, and the flavors of their cuisine? You might visit museums, walk down city streets, or browse country markets. You would definitely eat the food there, because that is one of the best and often most surprising ways to learn about a different place.

In some respects, the past is another country as well. It has flavors of its own that are well worth exploring. Experiencing those tastes reveals a time when the people and places of our own country were radically different than they are today. Like time travelers, we can see what life was like for our predecessors by conjuring up the techniques, textures, smells, and tastes of America from two hundred, three hundred, and even four hundred years ago.

The remarkable changes in ingredients, recipes, and menus over the centuries provide a window for us to appreciate just how different life has been during the various eras of America's story. Exploring our food heritage can also heighten our sense of the differences and similarities between then and now. For example, beaver tail is likely too gamey and fatty for our modern palates, but hungry fur trappers in the colonial period dined on it happily. Conversely, warm pumpkin pudding with heavy cream still appeals to us today, just as it enticed new colonists in seventeenth-century Massachusetts. So while this book is primarily about the evolution of America's national cuisine and "foodways"—a term that includes growing, harvesting, preserving, preparing, cooking, and eating food—history provides the context for understanding the intersection of culture and cuisine.

The American Plate provides a multilayered overview of the peopling of our country, our evolving foodways, and the transformation of our palates from 1400 to today. American Indians, Anglo-American women, enslaved (and free) African Americans, Chinese, Japanese, and Indian immigrants, western and eastern European arrivals, Hispanic families, and many more have contributed the flavors of their cultures to help create the variety of international and local influences we find in modern-day American restaurants, grocery stores, and kitchen tables.

To understand how America's diverse, edible heritage developed, it's worth exploring the people and events that shaped our cuisine and left indelible marks on the foods we savor today. Where did the raw ingredients for what we think of as American food originate? Were all foods local and seasonal before the introduction of modern transportation and preservation technologies? Who cooked the food, and whose culture predominated in the kitchen? These are just a few of the many questions we'll explore in these pages.

This book serves as a guide to introduce you to both an America of long ago and a more recent and familiar one, through more than one hundred different foods (or Bites) organized by ten eras of our national history. Drawing from a variety of sources, the book aims to shed light on the myriad cultures, values, and traditions that make up the United States through stories about our food (some very short, some longer). Peppered throughout are anecdotes, images, and recipes for all sorts of American dishes—from roast beaver tail and succotash to mint juleps, shoo fly pie, and firehouse chili.

Each chapter and each Bite stands alone, so you may read this book from front to back or just start with a time period or specific food that interests you. I chose certain foods because they and their history provide a particularly clear lens through which to view our broader national history. Others exemplify or are symbolic of a specific event, such as a version of a WPA soup recipe from the Great Depression. My hope is that these narratives and recipes will inspire readers to explore America's diverse culinary heritage, whether by recreating the unfamiliar tastes of the past or simply enjoying the stories I've included here.

Throughout America's history, nature, economics, technology, and immigration have played important roles in our adaptation of indigenous foods, development of new ones, and adoption of others from different continents. Ultimately, by learning about the origins of our richly diverse culinary heritage and exploring the history of the foodways included in this book, we can develop a deeper understanding of how we became the curious and passionate eaters we are today.

chapter 1

THE THREE SISTERS AND SO MUCH MORE

American Indian Foods before Columbus

 esty tomato sauce from Italy. Baked Irish potatoes, hot and comforting. Robust Indian curry with red pepper spiciness. You may think of them as originating in foreign countries, but these traditional dishes are actually all based on flavors from the New World, foods that traveled eastward from North America across the Atlantic in the hulls of Spanish ships more than five hundred years ago. They would revolutionize the way people ate around the world.

Less culturally defining, or perhaps just more routine, are the lowly beans and squashes that regularly appear on plates and in bowls around the world. These New World foods also changed diets, extending life expectancy and increasing population growth all over the globe, and while they may not have the zing of some of their more flavorful counterparts, they're equally important. And don't forget American corn, or maize, with its central role in much of American Indian culture. It is one of the most important food crops today.

The Americas have a remarkable variety of indigenous foods, and many foreign cuisines wouldn't look the same without them. South America gave us the potato in its various sizes and colors, which shaped the eating habits of northern Europeans—with devastating effect in nineteenth-century Ireland where the population had become too reliant on this one crop for sustenance. It's hard to imagine Italian cooking without tomatoes, which originated in

Mesoamerica (Central America) thousands of years ago, but there was a time when the future of pastas looked decidedly pale.

In the 1500s, Spanish conquistadors introduced tomatoes to Europeans, who eyed them skeptically. For one thing, tomatoes did not flourish in the damp, cool weather of northern Europe and Britain. Plus, their luscious appearance clearly labeled them as aphrodisiacs, while their leaves, so similar to their cousin, the deadly nightshade, linked them to poison. Fear trumped appetite, sexual or otherwise, so it is hardly surprising that the soft red fruit took a while to catch on. When it finally became clear that daring epicures did not die from eating what some people styled as "love apples," tomatoes flourished in the sunny kitchen gardens of southern Europe. Interestingly, both tomatoes and potatoes would travel back to the North American Atlantic Seaboard almost two hundred years later practically as novelties.

Other American Indian foods flourished in what today is the United States. These are the crops that many tribes grew, harvested, prepared, and bartered. The Three Sisters—corn, beans, and squash—and other food supplies made up the provisions that the American Indians generously shared with newly arrived British settlers along the Atlantic coast. The initial survival of the earliest colonies in Plymouth, Massachusetts, and Jamestown, Virginia, largely depended on the hospitality of the indigenous people with their food and cooking.

While you are reading the first ten Bites, it's important to remember that throughout North America, the First Nations had different cultures, cosmologies, and eating habits. Some were farmers, some were hunters, and some a mixture of both. Women in many regions gathered nuts, roots, berries, and wild greens such as watercress and fiddleheads, the curling tops of native ferns, in the early spring. In some tribes, women had also tended fields for generations, while others did not practice agriculture.

Note: In this chapter, I mention specific tribes or locations in an attempt to stay accurate, as broad generalizations are often unhelpful and inaccurate when describing American Indians before and after European contact.

BITE 1

Maize

(*Zea mays*)

People started farming in the Americas more than thirteen thousand years ago. In Mexico, archaeologists have found evidence of the cultivation of maize—what most Americans today call corn—since 7000 BC when ancient people domesticated and hybridized a wild grass called *teosinte*, the genetic ancestor of this versatile grain. Dispersed by wind, rain, and farmers sharing seeds hand to hand, the maize seeds traveled in all directions.

Over the centuries, native farmers selectively bred their crop to have larger cobs and bigger kernels, making the corn easier to harvest and process into food. As the cultivation of maize spread northward, different tribes developed various techniques and traditions for turning the hard seeds into nourishment. Maize became so central to tribal culture that its planting, sprouting, and harvest played important roles in religious observances and calendar reckoning.[1]

More than four thousand years ago, ancestors of the Hopi Indians were among the first indigenous people in the American Southwest to cultivate maize in what are now Arizona and New Mexico. It took about two to three thousand years more for maize farming to spread to the native tribes of New England, although some archaeologists believe that the cultivation of maize and other plants—including sunflowers and tobacco—happened independently on the East Coast.

Maize grew happily in semicleared fields without special plowing, which made it easy to cultivate. In many tribes, women tended the cornfields with their Stone Age tools, planting beans and squash around the low mounds where the maize grew. These three food crops—maize, beans, and squash—became known as the Three Sisters. The tendrils of the bean plant climbed up the cornstalk, supporting both the bean and corn plants, while the large flat leaves of the squash plants discouraged weeds. Today, anthropologists call this milpa agriculture, "milpa" being the ancient Nahuatl (Aztec) term for field.

The Hopi Indians in the American Southwest, called the "People of the Blue Corn," still grow a gorgeous, indigo-colored maize. You can buy it online to make authentically historical tortillas.[2]

Compared to some of the world's other domesticated grains, maize was an enormously productive crop that didn't require intensive labor. Wheat, for example, demanded more time and effort from the European peasant. Corn grew in poor or rich soils and happily shared space with other local crops as well as beans and squash. Once harvested and dried, the cobs or kernels could last all winter in covered pits or mounds. This was not the sweet, juicy yellow corn we buy today. The kernels were hard and variously colored—like the decorative Indian corn that stores sell now in the fall, only the cobs were smaller. Different kernel colors and cob sizes were identified with different localities.

Maize is a high-calorie carbohydrate and was an excellent food for the native people, who worked hard physically throughout the year. When the hard kernels were soaked overnight in an alkaline solution (such as water mixed with wood ashes), the heart of the seed was exposed and people could more easily absorb the nutritional value.

Corn is incomplete nutritionally, but when eaten with beans, it forms a complete protein. Adding bright orange squash to the meal contributes valuable vitamins. Thus, traditionally, the Three Sisters combined well not only in the field, but also in the bowl, making sturdy, nutritious dishes.

By the way, the word "corn" comes from an English term that refers to a region's local grain, which could be wheat, oats, or barley in England. The settlers called maize "corn," recognizing it as the common grain of the American Indians. Today—for better or for worse—corn and corn products are abundantly represented in our national diet and the global diet of many people as well.

NIXTAMALIZATION

About 3,500 years ago, native people in Mesoamerica developed a process called nixtamalization that improved the food value of maize. The word derives from an Aztec Nahuatl term for this treatment. They soaked the hard-shelled corn in water mixed with wood ashes or lime overnight. The softened hulls floated to the top of the water or were easily slipped off by hand.

Sweet, green corn, like we enjoy on the cob today, did not receive this treatment. But nixtamalizing the tough, dried kernels made their food value, including niacin, become much more accessible to the human gut. The Algonquin word for the resulting white, soft heart of the corn is *rockahominie*, from which our word "hominy" is derived. Once the outer hulls were removed from the maize, women could pound the hominy with a mortar and pestle or grind it on a stone by hand to make cornmeal.

Later, European settlers would skip this step in corn preparation because their millstones were powerful enough to turn corn into meal without soaking it. Unfortunately, that meant that their systems did not absorb all the nutrients from the maize. Thus, lacking nixtamalization as a culinary tradition, settlers with highly corn-dependent diets sometimes ended up with severe niacin deficiencies that caused diseases like pellagra.

Pellagra, which causes symptoms ranging from canker sores to memory loss, continued to be a scourge in poor farming areas in the South until the 1950s. American Indian and Mexican groups continued to soak their maize in the alkaline water, however, avoiding these problems.[3]

BITE 2

❧ Beans ❧

(*Phaseolus vulgaris*)

Like corn, beans have played an ancient, vital part in the traditional diet of American Indians, today and more than five hundred years ago. Most kidney-shaped beans—such as navy, scarlet runners, kidney, pinto, black, and lima—originated in the New World. Some details missing from the historical record can be more easily traced through word origins, and it's fun to see and hear the linguistic links with the past on our menus. For example, pronounce "lima" with a Spanish accent, *leema*, and you'll recognize that the lima bean was named after Lima, the capital of Peru, its homeland. The French word for green bean, *haricot*, comes from the Aztec *ayecotl*.[4]

> *As a little girl, I thought that pinto beans were eaten by pinto ponies, which had captured my imagination on TV. "Pinto" simply means "spotted" in Spanish, and both the beans and the ponies have similar spots.*

We often forget the humble bean's role as a change agent in the human story. Providing a cheap, reliable form of protein, it extended life expectancy among the Americas' First Nations and still provides vital nourishment to people all over the world. Following the cultivation of maize, the bean appeared as one of the Three Sisters in the common village plot set aside for farming. The black bean and the multicolored pinto bean are close relatives to the beans grown by pre-Columbian American Indians. Today, pinto and black beans abound in robust Mexican and vegetarian dishes.

Like maize, beans were dried and lasted well into the spring. They thickened the daily stews or soups, absorbing the smoky flavors of venison, buffalo, or salmon jerky and infusing the food with those deliciously complex tastes. Combined with corn, beans form a complete protein even in a meatless supper, which was crucial for indigenous tribes in the Americas when their game or fish supplies ran low due to heavy snows or other conditions.

Also, American Indians had essentially no dairy or domesticated poultry until the arrival of Europeans. Cheese, milk, and chicken eggs played no role as protein sources until the 1500s, when the Spanish conquered the American Southwest and introduced varieties of cows, pigs, sheep, and chickens. Thus, the combination of beans and maize in various dishes, like the traditional succotash recipe below, was of key nutritional importance.

TRADITIONAL SUCCOTASH

INGREDIENTS

- ½ cup water
- 1 (10-ounce) package frozen baby lima beans, defrosted
- 1 (10-ounce) package frozen corn, defrosted
- 4 to 5 smoked duck slices
- ¼ cup roasted sunflower seeds
- 1 pinch cayenne pepper
- Salt to taste

DIRECTIONS

Bring the water to a boil in a large saucepan with a cover. Add the beans and corn to the pot, and reduce to simmer. Cover and cook, 4 to 6 minutes. Drain. Meanwhile, dice and cook the smoked duck slices over low heat until crisp and all fat is rendered. Add the sunflower seeds, smoked duck crisps, and seasonings to the beans and corn. Combine well and serve.

Serves 6 to 8 as a side dish.

If smoked duck is unavailable or too pricey, substitute 4 slices hickory-smoked bacon. I like to add 2 tablespoons olive oil, but it is more authentic to add duck or bacon fat before serving.

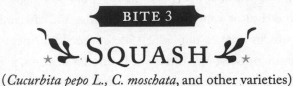

BITE 3

❧ SQUASH ✔

(*Cucurbita pepo L., C. moschata*, and other varieties)

The third and final of the Three Sisters, squash has always grown in a variety of shapes, colors, and sizes ranging from small pumpkins and acorn squash to thin, green zucchini types. There are two basic kinds of squash, summer and winter. Summer squash—like yellow squash and zucchini—have an edible, soft skin and soft seeds, and will last when ripe for two weeks in a cool, dry place. Winter squash—such as butternut, acorn, and spaghetti squash and pumpkin—have a hard rind and seeds with shells, and can be stored for months.

Although many of our squashes, like our beans, have been hybridized for taste, appearance, and shelf life, the traditional acorn squash is a good approximation of the plant grown by American Indians more than a thousand years ago among the cornstalks and the climbing runners of the beans. The drying process after harvest intensified the natural sugars of the thinly sliced orange squashes and pumpkins. This sweetness brightened the flavors of winter stews.[5]

Native women cleaned and dried the seeds of the pumpkin and other squashes, sometimes grinding them into flour. Local nuts—pecans, hickory, butternuts, black walnuts, and acorns, depending on the region—were also shelled and stored for winter, often by women and girls working together while they chatted and sang songs. The seeds and nuts brought texture, protein, and vitamins to the American Indians' diets. Stewed pumpkin, sweetened with dried berries and maple syrup, created a dish close to what modern Americans would call dessert.

It may be delicious, but don't confuse stewed pumpkin with pumpkin pie. The ingredients for a wheat-flour pie crust, the hen's eggs, the cane sugar, and most of the spices required to make a pumpkin pie would not be available on the North American continent until after the arrival of the colonists.

Once the women removed the soft insides and seeds from winter squash, the hollowed shells served as temporary eating bowls or containers. American Indians often ate one-pot meals, and both winter and summer squashes might be included in those big mixtures. The scooped-out shells of some squash varieties made pretty dishes to hold the stews, and cleanup was a breeze. Today, many kinds of squash are rising in popularity again, as people embrace everything from acorn squash and pumpkin to spaghetti squash as a healthier substitute for its pasta namesake.

BITE 4

VENISON

All this focus on corn, beans, and squash might indicate that American Indians ate a largely vegetarian diet. Far from it. Meat and fish of all kinds occupied a central role and took on spiritual aspects as well. The tribes who lived near the great natural cathedrals of early American forests developed efficient ways to hunt deer, a mammal that preferred cleared areas or the woods' edge to the forest itself. Native hunters drove herds of deer into rivers, where the animals could be easily snared by other members of the hunters' clan waiting for the prey. The men lit controlled fires driving deer into funnels created by high piles of brush and logs where small herds could be killed en masse. More typically, hunters expertly killed deer with weapons such as bow and arrows or spears.[6]

Venison functioned as a primary source of protein in the diet of many American Indian tribes, but the deer itself provided much more than just meat. Woodland Indians, such as many of the Algonquin tribes, used deer hide (with the hair) for winter clothing and as blankets. Expertly tanned or roughly scraped, the hide would become so central to American Indians' commerce with white settlers that a dollar's value became known as a "buck," or the price of a male deer hide. (Imagine what a million bucks worth of something back then would have looked like—that's a lot of deer hides!)

No part of the deer went unused. The skin itself, made soft and pliable by women's energetic pounding and scraping, could be used as a canvas for artwork or pictoglyphs, for housing, and for clothing. Tendons and muscles were stretched to provide webbing for snow shoes, traverse equipment, and papoose scaffolds. Antlers, bones, and teeth became weapons, tools, and ornaments. None of the carcass went to waste.

Although they were plentiful, deer were a challenge to kill with the Stone Age weapons at hand. Often hunters had to track a wounded deer for miles. If a hunting foray proved particularly successful, lucky and skillful men might return home with several dead deer or send others to fetch the carcasses where they were killed. After a communal feast, where everyone enjoyed the bountiful meat, women would set to work preserving the rest of the venison for the future.

WHAT DOES "STONE AGE" REALLY MEAN?

The term "Stone Age" does not necessarily indicate a time period. It refers to people who do not use or make iron or other metals outside of ornaments. In the early twentieth century, for example, some indigenous people in New Guinea and Brazil used Stone Age tools and weapons. Typically, Iron Age or Steel Age cultures ultimately overpower Stone Age cultures because their tools and weapons are stronger and often more effective. A war club is powerful until matched against a cannon. In the period between 1500 and 1700, Europeans armed with complex, metal tools and weapons conquered the First Nations in the New World. At the same time, European germs and plagues would prove to be equally devastating weapons of conquest against these indigenous, sometimes highly sophisticated cultures.[7]

Hunger frequently haunted the tribes in the late winter, when provisions set aside in the abundance of autumn began to run low and no food could be found on the plains or in the woods around them (depending on where they lived). Sometimes warring or rival tribes would destroy each other's food supplies to try to wipe out their opponents, or sometimes the winter would be unusually harsh and long—this was, after all, the Little Ice Age. Some Algonquin tribes referred to February as the Hunger Moon.

Most native people preserved meat during the hunting season by hanging it in strips above a slow fire to create jerky. (Our word "jerky" comes from the Peruvian Quechua *ch'arki*, meaning smoked or burned meat.) Any meat could be preserved this way, including bear, duck, turkey, buffalo, and rabbit. Women would remove the fat and melt it slowly, rendering it separately to extend its freshness.

> *Tribes preferred fatty meat over the lean meats recommended today for modern diets. Bear fat, incidentally, turns rancid rapidly, but once rendered it was valued as a strongly flavored cooking grease, a healing lubricant for wounds, and a cosmetic preparation for skin and hair. Rendered venison fat served as an important and desirable ingredient in many Woodland Indian meals.*

BITE 5

⚘ AMERICAN BISON ⚘
(*Bison bison*—also known as the American Buffalo)

Deer weren't the only four-legged creatures that American Indian tribes feasted on in early days. Vast herds of bison inhabited most of today's central continental United States, stretching from the western slopes of the Appalachian chain to the Rocky Mountains, in forests and on the prairies. Smaller herds existed along the East Coast but died out soon after the Europeans' arrival.

A bison bull is a huge mammal, weighing in at about a ton, while the bison cow averages seven hundred pounds, dainty by comparison to her mate. The Plains Indians, a category that includes six language groups and many tribal nations—Oglala and Lakota Sioux, Crow, Blackfoot, Comanche, and others—identify themselves as people of the bison, because their life, culture, and ultimately, their destiny were inextricably tied to these great, shaggy beasts.

A classic image of American Indians hunting bison shows them riding expertly on horseback while shooting arrows at the herd. But there were no horses in the Americas until the arrival of the Spanish in 1492. So, before that, for more than two thousand years, men hunted buffalo on foot, sometimes chasing herds over cliffs for a mass slaughter.

This technique of herding bison to their death proved extremely effective, and a clan could feast for days on the choicest cuts of the bison, such as the tongue and liver.[8]

In what is modern-day Montana, there are more than three hundred buffalo jump archaeological sites, including one, now the First Peoples Buffalo Jump State Park, where eighteen feet of ancient buffalo remains lie at the foot of the cliff.

So many millions of bison roamed the Great Plains that it was unfathomable that they would ever near extinction. Once the American Indians

captured wild horses, descendants of the animals imported by the Spanish in the 1500s, they became among the most accomplished riders in history. Hunting on horseback increased their chances of bringing home large game for their families. Still, they barely made a dent in the bison population. It would take bovine disease from domestic cattle, the railroad, and a stated federal policy of buffalo slaughter designed to force starving Indians onto reservations to bring these shaggy giants to the edge of oblivion in the 1890s.[9]

Like deer, a slaughtered bison served multiple purposes for the Plains Indians, providing food, clothing, shelter, weapons, ornaments, and tools. A buffalo robe or blanket kept its owner warm on the coldest winter nights. Women smoked dried strips of buffalo meat and rendered the fat to store for the lean winter months. Bison meat was a key staple in the life of the Great Plains Indians until the 1890s.

Bison meat is very nutritious compared to that of domestic animals. It ranks higher in iron and vitamin B-12 than beef, pork, or chicken, and ranks lower in fat, calories, and cholesterol. The fat is stored separately on the back of the bison and does not marble the meat. A plain, well-done bison burger can be very dry, at least to modern tastes, so don't overcook it.

PEMMICAN

Pemmican is a densely nutritious, portable food with an ancient history. It is the American Indian version of an energy bar. Pemmican traditionally includes some kind of meat and animal fat. So you can also make this from dried venison, duck, moose, or turkey, and it will still be authentic.

INGREDIENTS

- ► 1 pound natural (unflavored) buffalo jerky, finely minced
- ► 1 cup dried blueberries
- ► ½ cup toasted hickory nuts, pecans, or sunflower seeds, finely chopped
- ► 2 to 3 tablespoons warm buffalo fat

DIRECTIONS

Mix the buffalo jerky, blueberries, and nuts together thoroughly. Using a pastry cutter or two knives, slowly cut in 1 tablespoon of soft fat at a time, stirring after each addition. Test the mixture after each addition, and stop adding fat once the mixture stays together in a clump. Spread mixture in a rectangular pan, 11 by 7 inches. Slice into bars when cool. Place in individually decorated rabbit-skin pouches, or wrap in waxed paper. Store in a cool, dry place. Take along on hunting trips or anytime you are going on long journeys on foot.

Makes 1 dozen bars.

Substituting venison fat for the buffalo fat would also be authentic, or try using beef suet instead. Pemmican is good for you if your life requires that you burn enormous amounts of calories. Otherwise, the fat is detrimental to your health.

BITE 6

❧ BLUEBERRIES ❧

The humble blueberry is experiencing a big revival these days, thanks to publicity about its antioxidant qualities and low carbohydrate value. But for many people—including me—blueberries have always been in style. They grow all over North America except in arid regions, so it's easy to take advantage of them. I spend my summer vacations near the Garden River First Nation Reserve on St. Mary's River in Ontario, and the First Nation people there sell baskets filled with tiny blueberries, intensely sweet and flavorful. Enjoying a breakfast of wild blueberry pancakes and local maple syrup while sitting outside on our front porch defines summer for me.

Raspberries, huckleberries, and blackberries, all of which still grow wild in the continental United States, join blueberries as some of the most delicious and popular indigenous small fruits of North America. They played an important role in taste and nutrition, particularly among the Woodland Indians of the East, the Coastal Indians of the Pacific Northwest, and the tribes around the Great Lakes region, where they were particularly prolific.

Some tribes, for example, were great hunters in the North Woods and spent little time farming, although they traded with other tribes for corn and beans. They did, however, gather blueberries, eating them fresh and drying them to keep for the winter season. Although American Indians in the dry Southwest brightened their cooking with chili peppers, food in the North could be bland. Though it may seem weird to our modern palates, berries enlivened a mixture of corn and beans considerably for many tribes.

> *Humans have been mixing fruit with meat for millennia. The North African tagine today is reminiscent of a medieval European spiced stew with fruit, a popular dish until the eighteenth century.*

Traditionally, American Indians across the continent enriched their winter repasts by adding dried blueberries and other wild fruit. And women added blueberries into pemmican, the American Indian version of a portable protein bar.

Today we consume blueberries as a healthy snack, in yogurt, and in baked goods such as pies and muffins. But we generally keep them away from meat-centric recipes. An elegant presentation of duck breasts might include a warm blueberry sauce, but in general—except for applesauce—we no longer mix fruit regularly into our meat dishes. Too bad for us!

BITE 7

⚘ MAPLE SYRUP ⚘

Imagine life today with no sugar or honey. I don't mean only occasional, judicious amounts of sugar, maybe a slice of cake on a birthday, for example. I mean none. I think there might be protests in the streets! We humans have always had big sweet tooths, and like people today, early American Indians loved sweet tastes. But they had no access to cane sugar or honey. So they invented maple syrup, a truly ancient food that we still encounter in our truly modern lives.

The expensive amber syrup we pour on our pancakes is the concentrated sap of the indigenous sugar maple (*Acer saccharum*) and its relatives that grow in the northeast and upper midwest woodlands of today's United States. Food historians believe that American Indians used maple sugar to enhance their cuisine much the way contemporary Europeans added salt to make their food more palatable.

Some natives called March the Maple Moon, after the time when the spring sap begins to run in the sugar maples. Long before white men arrived

A SWEET AND BITTER STORY

Maple sugar is just one kind of highly concentrated natural sugar from the Americas. In the Southwest, Indians processed the agave cactus centuries before the arrival of the Spanish, producing another natural form of sweetener, agave syrup, that is growing in popularity today.

White or brown sugar from sugarcane, a woody plant imported by white settlers to the Caribbean, produced enormous riches for European sugar-plantation owners there and eventually in Louisiana as well. They used a particularly brutal system of plantation slavery, with a very high mortality rate, to bring the desirable luxury product—called cane sugar, or simply sugar—to the European and American table. White cane sugar has a neutral taste, while maple sugar has a specific identifying flavor which made it less valuable as a commodity from the British and European point of view. And unlike cane sugar, it has a limited shelf life.

While sugarcane is an annual crop with a predictable productivity, sugar maples take years to grow to maturity. Mass production remains challenging, even with today's modern technology, and maple syrup is well regulated to guarantee a premium result.

on the continent, Ojibway Indians suspended hide or birch-bark buckets underneath small deep holes they had bored into the trees, and the sap trickled into the buckets as winter turned into spring. The thin sap was then heated in wide moose-skin containers. Hot stones from the hearth were placed in the containers, bringing the sap to a boil and condensing it into syrup.

In other communities, the sap would be allowed to freeze overnight, and the ice formed at the top would be removed each morning, ultimately leaving a thick syrup in the bucket. Ojibway women would pour the light amber, intensely sweet liquid into large birch-bark vats, or scrape up the crystal residue and use the brown maple sugar to sweeten fats and meats.[10] Placing hot rocks in and fishing cooler rocks out of large vats of hot, sugary liquid is tremendously messy, tiring, and sometimes dangerous work.

When you read the descriptions of the painstaking efforts required to boil and condense the maple sap in hide or bark vats, you understand why American Indians would so highly value the iron pots that Europeans eventually introduced as a trade item. Iron pots made syrup production faster and easier for American Indians, a welcome step in technology brought from Europe to the New World.

But even with today's state-of-the-art processes and equipment, the extraction and production of real maple syrup is still not an easy or cheap process. So next time you're enjoying a stack of pancakes drizzled in that amber golden syrup, give a silent thanks to our indigenous forbearers.

BITE 8

WILD RICE

(*Zizania palustris*, *Z. aquatica*, and *Z. texana*)

Like maple syrup, wild rice has a growing fan base among Americans who want to eat unadulterated, simply processed foods. It is delicious in baked goods, as a hot cereal, as a side dish, or as an entrée paired with other ingredients. Also, like maple syrup, wild rice is unique to the New World.

European settlers knew that Asian-type rice grew in shallow water paddies. Rice had been grown in Italy since the late Middle Ages when it was imported from the Far East. When they witnessed native tribes harvesting grain along the shores of the Chesapeake Bay and the pristine Great Lakes, explorers assumed these narrow black seeds were also a type of rice. (Actually both Asian rice and *Zizania* are types of cereal grass, along with wheat, oats, and barley.) American Indians collected wild rice in their canoes by carefully knocking the seeds from the top of the stalks onto the floor of their boats.

Tribes gathered together to celebrate the wild rice harvest, usually around mid-August. The Ojibway called this time of the year the Wild Rice Moon. The rice was dried on flat rocks in the sun. Ojibway historian Lolita Taylor writes, "Nothing can equal the aroma of the ricing camp…A contented feeling of well-being filled the camp. The first grain of the season had been offered for a blessing from the Great Spirit. Boiled with venison or with ducks or rice hens, it was nourishing and delicious."[11]

Particularly in the Great Lakes region, wild rice served as an important food staple. Often women added it to the stew pot, contributing flavor and substance to the simmering collection of foodstuffs. Dried, it could be ground into a dark flour to make a flat bread. Served with chopped nuts,

American Indians harvesting the cereal grain known as wild rice.

fresh blueberries, and a bit of maple syrup or maple sugar, it was a nutritious and delicious hot cereal, very similar to a hot cereal modern Americans would enjoy today. Of course, the Ojibway might have added bits of fish or meat if they had some on hand, because they comfortably mixed sweet and savory in one dish more frequently than we do in the twenty-first century.[12]

BITE 9
❧ RED PEPPERS ❦
(*Capsicum*)

While wild rice and maple trees flourished in the northern part of what is now the United States, very different kinds of crops thrived in hot, dry regions. In the American Southwest, native inhabitants seasoned their food with chili peppers whose seeds had followed ancient trade routes from Mexico and Central America northward across the Rio Grande. Varieties of *Capsicum* (red peppers) grow wild in South America, and archaeologists have found traces of cultivated peppers from 2000 BC.

From sweet red bells to explosively spicy habaneros, red peppers require a long, hot growing season. *Capsicum* peppers boast not only a high vitamin C quotient but an antimicrobial agent as well, helping to keep food from spoiling in warm climates. While some peppers dry well in the sun, indigenous people found that other variants benefited from being smoked over a low fire. Roasted, smoked, and ground jalapeño peppers, for example, produced delicious chipotle powder, one of my favorite flavors in regional American cuisine.

> *Historically, northern writers accustomed to milder flavors denigrated the use of hot spices by suggesting that the intensity was used to mask bad meat or other rancid ingredients. Today, scientists have proved that hot Capsicum varieties actually help to preserve food and can inhibit bacterial growth microbiologically.[13]*

Incidentally, Christopher Columbus caused the world a lot of confusion by referring to these sweet and spicy fruits as "peppers." In fact, they are not related in the least to the small, round, black spice that helped inspire his voyage of discovery of the New World, but that didn't matter to him. A marketer as well as an explorer, Columbus knew his royal Spanish investors were eager for their own source of the highly expensive black dried seeds called pepper, imported overland at great cost from Asia. He brought *Capsicum* peppers back to Europe, where they became popular in the hot, sunny climates that replicated their native habitat.[14]

A drawing of the Capsicum *(red pepper) plant.*

Hot and sunny, however, does not describe Britain, especially during the chilly seventeenth and eighteenth centuries, when so many English and Scottish colonists emigrated to North America. Unaccustomed to red pepper, these Anglo-Americans often preferred the gentle flavors of their herb gardens. While Spanish and Moroccan gastronomy embraced the New World *Capsicum*, the white settlers along America's Eastern Seaboard were not so welcoming. These food traditions continued for generations, and chili peppers retained their exoticism among some Yankees until the mid-twentieth century and even later.

Traveling through the Southwest, you may frequently spy a braid of dried red chilis, (*ristra* in Spanish, meaning rope), adding a dash of bright color to local doors and kitchens. Actually, ristras are available online and in catalogs all over the country these days. The dried chilis are not just vivid; they symbolize good luck—and foretell a spicy dinner.

Today, their hot spice flavors a variety of foods popular from coast to coast, northeast to southwest. Whether featured in Texas chili or Thai curry, *Capsicum* finally plays a central role in food across the United States.

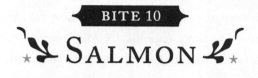

BITE 10

SALMON

Perhaps by now you've realized that American Indians enjoyed many ingredients that might be featured on a modern gourmet's short list of fine foods. Salmon is no exception. Like venison, salmon was not unique to the New World, but it was and continues to be one of the great gastronomic delights of the world.

Prior to the arrival of Europeans, the waters of early North America literally teemed with fish. American Indians harvested them with nets, lures, weirs (fish traps), and spears. In the spring, the Hudson River would boil with migrating shad, a huge bony fish that was cooked on soaked cedar planks. Cleaned and gutted, fish could be consumed immediately, added to the daily stew pot, or roasted on green sticks. It also could be dried or smoked for future use.

Different tribes ate only the fresh fish that swam in their local waters, since dead fish rots so quickly if it is not preserved. The Powhatans on the Chesapeake feasted on the abundant seafood of the great bay, enjoying soft-shell crabs in the spring and oysters in the fall. The Shinnecock and Montauk tribes of Long Island bravely hunted whales in oceangoing canoes and eventually shared their remarkable skills with Anglo-American whalers.

Along the Atlantic and Pacific coasts, a beached whale could feed a clan and provide an assortment of resources, from blubber to multipurpose whalebone. It is easy to understand why indigenous people interpreted a beached whale as a spiritual and physical gift from the great Creator, like leviathan manna from heaven.

The European tradition of saline preservation of food—salt pork or salt cod—was not as common among American Indians, since most didn't have steady access to salt. Smoked fish created a concentrated, low-fat form of protein that kept well throughout the winter without salt. Today we consider smoked salmon an epicurean treat.

The Creator also sent the salmon to the Pacific Northwest, according to the beliefs of the numerous tribes that revered the great fish with pink meat. In the northwestern cultures, chinook (or king) salmon occupied a central place, and its harvest during its annual migrations to spawning grounds was accompanied by hard work and solemn thanksgivings as well as feasts.

The Yakima from today's Washington State, for example, had rich folklore around the salmon and its habits. In one story, their Creator specifically instructs the Yakima not to take more salmon than they need. Scholars estimate that, in fact, native people harvested the salmon for at least 1,500 years without upsetting the local ecosystem. Needless to say, it took only a few generations of European settlers to seriously damage the salmon population in the Columbia watershed.[15]

Just like the people of the buffalo and of the deer, the people of the salmon generally divided the provisioning tasks along gender lines. The men crafted cunning harpoons and large dip nets, along with weirs woven with reeds, to trap the chinook and other fish, some of which weighed close to one hundred pounds. The women took charge of the cleaning and preserving, hanging strips of salmon meat over outdoor racks to air-dry or smoking the strips within very small smoke huts that belonged to individual families. They used slow, green alderwood fires to impart a sweet smokiness to the pink meat.

Today, salmon remains extraordinarily popular. Some of the salmon on the market is raised on fish farms, which lowers the price and reduces the stress on the wild salmon population, but raises pollution and crossbreeding concerns. Wild salmon is making a comeback in areas from which it had disappeared for dozens of years. During my summer vacation on St. Mary's River, I now see the salmon jumping on their way toward Lake Superior, but in the 1960s my grandmother thought they were gone forever, victims of pollution, habitat destruction, and overfishing. Today it is a joyful sight to see one leap into the air, bursting out of the water in a startling arc of silver.

Wild salmon still need our respect, care, and oversight. American Indians fished for salmon for millennia, never impacting the population of these wild creatures. As we will see in the next chapter, eventually the European settlers would challenge this abundance.

chapter 2

COD, BEAVER TAIL, AND SASSAFRAS

Encounters and Exchanges, Old World and New

any fifth graders know that the Vikings were the earliest Europeans to reach North America, but the Spanish were the ones who arrived and stayed, almost seven hundred years after their Nordic forerunners. In 1492, Italian-born Christopher Columbus sailed under the flag of his Spanish venture capitalists, Queen Isabella and King Ferdinand, having lured them into funding the venture with promises of returning home with ships loaded with riches and spices from the East.

Columbus's voyages turned out to be an enormously profitable investment for the Spanish crown, and an inspiration for other explorers. Spanish conquistadors and Jesuits followed in Columbus's wake, looking for rich lands and desperate souls to conquer or convert now that the heathen Moors had been driven out of Spain. Sons of families with a military tradition sought new opportunities in the armies of the conquistadors.

The settling of America's East Coast by the English gets a lot of attention in history books. But one hundred years before the *Mayflower* unloaded its seasick passengers in Massachusetts, Spanish colonists settled what is now Florida and New Mexico, bringing their own food traditions and adopting (and adapting) American Indian ones as well. The Spanish exported the first maize, turkeys, potatoes, chocolate, various beans, squash, and tomatoes to Europe, where their use spread slowly but surely across the continent, east to Asia, and south to Africa.

Some Christian leaders frowned on the novelty of these foods. After all, they weren't mentioned in the Bible. But the ease with which some plants grew—witness the potato in Ireland—encouraged wide adoption by farmers eager to feed their hungry families and livestock. Many of these foods still play central roles in cuisines around the world today.

New World foods faced a mixed reception. Peasant farmers generally avoided taking risks with their harvests and often resisted new crops. In fact, for centuries, many Europeans disdained maize as a grain suitable only for feeding farm animals. Italian peasants, however, reinvented cornmeal mush as luscious polenta, possibly because maize was not listed in the traditional tariff laws and so went untaxed. Anything that escaped the tax man had a profound allure, then as now. Tomato sauce, that mainstay of the Italian kitchen, became popular in the 1700s, which qualifies as the dawn of the modern food era in Italian reckoning.

While the Spanish introduced many foods from the Americas to Europe, Asia, and Africa, an even wider range of edibles arrived in North America *from* Europe, first with the Spanish and later with the English colonists. Remember, American Indians in what is now the United States had no domestic animals other than the dog, for them a truly multipurpose mammal that could hunt, guard, carry, and be eaten when times got tough.

Europeans brought livestock—poultry, cattle, horses, and sheep—which literally recreated the landscape, the work patterns, and the diet of America. The new settlers' introduction of cattle provided beef, dairy, and the sturdiest of North American draft animals, oxen. European colonists understood cattle as vital to their transplanted agricultural economy.

Through their foraging and overgrazing, however, the cattle had an immediate negative impact on the local environment, as did the horses, hogs, sheep, and goats that were also brought over in ships' holds. In 1680, during an early rebellion against the Spanish in what is now New Mexico, the Pueblo Indians released the local Spanish livestock and drove the animals away from the settlement. The horses, hogs, and sheep multiplied in the wild, becoming part of what appeared to be the indigenous American landscape.

But even with this introduction of new food sources along the Eastern Seaboard, new arrivals frequently suffered through a starving time during the first year of settlement. Generally, they did not transport enough food to hold them until the completion of the next year's growing season, a fact few of them took into account before boarding their ships.

Jamestown, Virginia, founded in 1607, was the first permanent British settlement in North America, but it barely escaped total failure due to lack of food. Inadequate planning and ignorance, combined with the worst drought in seven centuries, created conditions promising only for the Grim Reaper. Disease and famine wiped out about 85 percent of the English population during the winter of 1609–1610. Only the arrival of supply ships from England—bearing provisions, more settlers, and new leadership—saved Jamestown from abandonment.

Jamestown wasn't the only near-casualty due to serious food shortages. In New England, half of the passengers and crew from the *Mayflower* died from exposure and malnourishment during that first winter on Cape Cod in 1621. But the generosity of the Wampanoag people helped the rest struggle through, and rapid adaption of Old World farming to the demands of the New World produced abundant food within the first year for those colonists. Today we refer to them as the Pilgrims, a label fastened on them by nineteenth-century admirers. These surviving colonists celebrated their first successful harvest with a feast day now known as Thanksgiving.

Once the starving times ended, at different times for different colonies, the settlements by and large prospered. Some families began to accumulate wealth, while a few remained poor. Putting food on the table still required a lot of hard work by farm owners, servants, slaves, and children. But for most white people in early America, food was abundant and cheap by European standards.

By the late 1600s, the Eastern Seaboard moved from a subsistence

One of the foods that saved the Plymouth settlers was maize, which they stole just after they landed on Cape Cod from the local Wampanoags. Despite the theft, the American Indians shared more maize when they realized how close to starvation the foreigners were. And in the spring of 1621, an English-speaking Indian named Tisquantom or "Squanto" taught the survivors how to grow the Three Sisters. Maize was now a staple, not only for indigenous people, but for Anglo-Americans as well.

agricultural society to a market economy, exporting foodstuffs, tobacco, timber, and furs across the Atlantic and down to the Caribbean Islands. And with the discovery and use of these new foods and products, the inklings of American cuisine began to emerge.

BITE 11

❧ Jamaican Pepper ❧
or Allspice

(*Pimenta dioica*)

When Columbus set off from Spain in 1492, he was in search of a shortcut to India, the source of a highly priced luxury item for which rich Europeans had clamored for centuries: pepper. Originally, black pepper arrived on the European continent after traveling along old trade routes from India to Persia to Constantinople. With each leg of the journey, the price of the pepper increased dramatically. But the ancient Romans craved the sharp bite of pepper to enliven their diets, and the wealthy were willing to pay the market price.

The spice continued its popularity throughout the Middle Ages, despite (and perhaps because of) its exorbitant cost keeping it out of the hands of the average citizen. For much of that time, it was expensive but not impossibly so. That is, until 1453, when Constantinople fell to the Ottoman Turks. The new rulers of the spice routes slapped a huge tax on pepper and its price skyrocketed.

As we have seen, Columbus was the canny combination of a visionary explorer and a stellar marketer. So he spied an opportunity in the mass panic over this precious spice and convinced Queen Isabella and King Ferdinand that he knew how to get pepper at wholesale prices. Best of all, his method promised to avoid those darned Turks entirely. Believing his analysis that his ships would inevitably run into Asia if they just sailed long enough, the king and queen funded Columbus's trip westward across the Atlantic from Spain. And he would have reached Asia's Pacific shores, except for one small problem: the Americas got in the way.

Columbus believed that he had reached the east coast of India—thus naming the native peoples "Indians"—a confusing inaccuracy for American Indians that persists five hundred years later. He searched fruitlessly for the esteemed black pepper to ship home to Spain, hoping it would make him and his adopted country fabulously wealthy. But the Caribbean Islands offered another type of spice, which the Spanish explorer cleverly christened "Jamaican pepper." It might not be the black pepper of India, but it was pretty close. And India had to be nearby, somewhere!

WOULD THE REAL PEPPER PLEASE STAND UP?

Columbus, of course, never made it to India, but he never gave up on finding pepper. By the way, *Piper nigrum* (black, white, or green peppercorns, which are all from the same tree) and the *Capsicum* families (red peppers) are not related. Not even remotely. It was partly a marketing ploy to sell the idea of red peppers on the European continent and partly Columbus's conviction that the Americas were *really, really* part of Asia.

Red peppers rapidly became popular, especially around the Mediterranean basin, where they thrived in the hot summers. Black pepper would eventually decline in price as European explorers ventured around Africa and across the Indian Ocean to India, the original source. While still a luxury flavoring, it no longer commanded a crazy-high price.

Today we call Jamaican pepper "allspice" because to the people in the late Renaissance who christened it that, this powerful spice seemed like a combination of cloves, cinnamon, and nutmeg all in one dark kernel. Today, we recognize its flavor primarily from pumpkin pies and other spiced baked treats often associated with Thanksgiving and Christmas.

Without the traditional holiday recipes passed down for generations, perhaps we might have forgotten about allspice. Except for one thing: that wonderful Jamaican technique for preparing meat dishes called "jerking." The word comes from the same Quechan root as jerky, and simply means preserved or smoked. Allspice is a key ingredient of Jamaican jerked pork and chicken.

Jamaican pepper or allspice held a prominent position in early American spice cabinets, enhancing everything from hot drinks to chicken stew to puddings. Eventually, its popularity waned, the way food trends, like fashion trends, come and go. Allspice continues to play a starring role in jerked meats and holiday baking.

BITE 12
ATLANTIC COD
(*Gadus morhua*)

Columbus hoped to strike it rich with spices, but a true fortune swam in the waters of the New World. For thousands of years, the Northwest Atlantic teamed with codfish. That opportunity may have sailed by Columbus, but it wasn't lost on other adventurers. In the 900s, Vikings sailed to the Grand Banks, off the coast of Labrador, and returned with ships laden with great fish. Centuries passed before other Europeans followed, casting vast nets into the sea and bringing up masses of huge *Gadus morhua*—the Atlantic cod, many over six feet long. Gutted, split, salted, and dried, the preserved cod could be stacked like pale boards and shipped back to the continent. The stiff, white fish stayed edible for months, if kept dry.[1]

COD BRINGS THE FISHERMEN, THE FISHERMEN BRING PLAGUES

Occasionally, fishermen-turned-explorers would sail west—seeking freshwater, blown off course, or simply inquisitive—and land in northern New England. They returned home with tales of vast forests and glimpses of strong, tattooed inhabitants. In the early 1600s, around the time of William Shakespeare, some of these fishermen-explorers kidnapped American Indians and brought them back to Britain as curiosities. Others departed with no captives, but left behind the germs of common European diseases to which the indigenous population had no resistance.

Between 1608 and 1610, a plague swept from Maine down through New England, decimating American Indian villages. Entire clans were wiped out and villages abandoned. Microscopic germs had brought down entire civilizations with the help of the Spanish conquistadors in Mexico and Peru almost a century earlier. Now the pattern repeated itself along the North Atlantic coast.

Before the arrival of the *Mayflower* in 1620, Old World diseases introduced by fishermen had already cut like scythes through the natives on Cape Cod. This made settlement much easier for the early English colonists. In fact, they believed it was a sign that God had prepared the land before them. Providence was on their side.

Of course, the cod had no idea about Providence or their role in enticing Europeans across the Atlantic armed with fishing nets and deadly germs. But their abundant schools off the coast of North America changed the course of history.

Cod proved an economic boon for the British settlers who colonized New England. Since fishermen from Massachusetts soon exported salt cod by the ton, a carving of the famous fish hung from the Boston statehouse to remind all passersby, as well as government officials, of the importance of fishing to that colony's economy.

Not just an export, salt cod served as a mainstay of the colonial diet up and down the coast, and inland as well. Like a well-smoked ham, but cheaper, it didn't go bad for months. A housewife or cook chopped off quarters of the broad, flat splits and soaked them in water. The rest could be saved for another day. The water might need to be changed two or three times to get rid of the intense salt.[2]

Now soft and somewhat reconstituted, the moist cod could be mixed with other ingredients to form cod cakes, a common and beloved specialty served at any mealtime. Some liked cod cakes at breakfast, or at supper with eggs or perhaps some boiled spinach, or alongside other dishes for their midday dinner. They would have consumed their hearty cod cakes with a spoon and a knife. During the 1600s and most of the 1700s, average colonists did not use forks for eating.[3] By this time, however, the cod was way past caring about dining etiquette.

We may choose to eat cod with a fork and knife, chopsticks, or on a roll today. Perhaps we should not eat it often, however. Due to chronic overfishing in the late twentieth century, the Atlantic cod is close to commercial collapse. Government quotas are helping to restore the cod population for the long term, a process that will take years. But it is doubtful that fishermen will ever see those six-footers in their nets again.

BITE 13

PORK

Like cod, pork provided a reliable source of protein for the early settlers. Columbus first introduced pigs to the New World in 1493. More than one hundred years later, the first English settlers brought pigs on board their supply ships to the East Coast. The pig was a very familiar animal in the Old World. Peasants all over Europe raised hogs. (As livestock goes, they are relatively undemanding.) But in the New World, hogs found paradise. They ran through the forest and feasted on debris. They required little time, expertise, or care by their owners. This trait proved particularly important in early Virginia, because the colonists there were too busy raising tobacco to have to worry about pig farming.

Pigs happily gorged on wild acorns under the vast oaks of seventeenth-century Virginia. Slaughtered in the late fall, when the frost would keep the meat from rotting, pork was smoked and hung, or salted and packed into barrels. Virginian settlers used ham to add flavor and nourishment to an array of foods. Fatty, sometimes moldy salt pork was eaten by the poor, and small amounts were distributed to servants and slaves.

TOBACCO: "NOXIOUS WEED" AND BROWN TREASURE

John Rolfe—the same John Rolfe who would marry Pocahontas—planted the first tobacco seeds in the Virginia colony and harvested a new type of treasure. Tobacco quickly became a cash crop, so profitable that most other farming was neglected. The labor-intensive "noxious weed" required cleared fields and many hands before it could be shipped across the Atlantic, where it fetched eye-popping prices.

The early settlers in Jamestown, Virginia, arrived ill-equipped for agriculture and quickly alienated the Powhatan Indians who had thrived along the Virginia coast, practicing milpa agriculture and enjoying the stunningly rich seafood supplies of the Chesapeake area. The English hoped to find vast mineral wealth like the Aztec and Incan plunder of their Spanish rivals. But soon the fortunes of the Virginia Company looked grim indeed on that front.

There were no mega mineral lodes to be found in the Chesapeake region. Poorly nourished and sickened by bad water, the Englishmen accomplished little more than causing trouble. Then in 1614, Rolfe exported the first four barrels filled with the large brown leaves of tobacco, which sold for a fat profit in London and inflamed the hopes of the settlers. Fortunately for these men (and they were mostly men), maize and pork needed much less work than that demanding but extremely profitable cash crop.

Early Virginia, however, did not have a monopoly on hog raising, nor were those settlers the only ones who relied on it heavily in their diets. Although householders throughout the colonies might prefer beef, they consumed pork of all kinds more frequently because it was cheaper.[4] Throughout the colonial period, enterprising merchants in the small but growing towns sold salt pork, and salt beef as well. As early as 1675, market-oriented farmers on Long Island, New York, for example, packed salt pork in huge barrels and shipped it to the Caribbean sugar plantations.

There, plantations were so focused on producing their sugar that they did not raise enough food for their enslaved workers. So plantation managers on the sugar islands, like Barbados and Jamaica, purchased the cheap salt pork from the North American colonies to feed their largely African workforce. One can only imagine the quality of the meat by the time it arrived in those subtropical ports. Don't even try to imagine the smell when one of those barrels was opened.

Today, Virginia is justly famous for its dry-cured country hams, continuing a culinary heritage that began in the seventeenth century. Perhaps the most famous brand is a Smithfield ham. Surryano ham, from Surry County, Virginia, is a relative newcomer and well worth seeking out. Traditionally, thin, flavorful slices of smoked ham are served inside a warm biscuit.

The famous writer and wit Dorothy Parker once defined eternity as "two people and a ham." A big smoked ham does seem to last forever in a small household. But one quarter cup of diced smoked ham adds just the right oomph to a multitude of recipes, so forever is okay by me.

BITE 14
BEAVER TAIL

News flash! Today, Americans no longer consider beaver tail a desirable food. It's the type of thing you eat to show that you can, sort of like eating live bugs. Beaver tail is essentially gamey-tasting fat, with swampy overtones.

However, it is not complicated to prepare if you have an open fire at hand. Then again, chances are, an open fire is not a challenge for you if you're planning on eating beaver tail.

Maybe this colonial delicacy doesn't quite tantalize our modern taste buds. But it's worth exploring briefly because beaver tail suited two types of people in early America perfectly. And without these people, we wouldn't be where we are today. The first were American Indians, who ate a lean diet (especially when you consider how many calories they burned every day due to their rigorous lives) and could benefit from the caloric burst of an occasional beaver tail. For example, a fatty beaver tail could be very nutritious for an indigenous family during the Hunger Moon, that time in late winter when food supplies were meager.

Secondly, a roast beaver tail could be the ideal supper for a fur trapper in the winter wilderness. (Fur trappers could be English, Dutch, French, or American Indian.) The plus side? Having beaver tail to eat meant that you had beaver pelts to bring back to the trading posts. And beaver pelts were the most sought-after fur from the sixteenth century into the nineteenth.

Beaver wasn't just for fur coats or collars. Hatters used sheared beaver pelts to produce the best glossy black hats. And the market demand for this fur was enormous. In fact, fur trappers constantly pushed westward in search of more beavers, because they had trapped out many parts of the East Coast. The fur trade proved enormously important in opening up the interior of North America to exploration. Settlers and their families were not far behind, turning forests into fields and orchards.

So a trapper with beaver tails to roast was a lucky man indeed. He might have been cold, isolated, and hungry. But he had enough pelts to make some good money, and he had a rich, fatty slab to roast over a fire for his supper. And hunger makes the best sauce.

ROAST BEAVER TAIL

You will need an open fire to cook the beaver tail, which is covered with a rough, scaly skin. If you happen to have a beaver tail on hand, I am sure that an open fire is no problem for you.

INGREDIENTS

▶ 1 beaver tail, well washed
▶ Salt and freshly ground black pepper to taste

DIRECTIONS

Use long-handled forks or skewers to hold the tail over the fire until the skin blisters all over both sides. Remove the tail from the heat, allow it to cool, and then peel the skin off. Roast the remaining fatty slab over glowing coals until thoroughly cooked, using a long-handled fork or a long-handled metal tool for cooking hamburgers over coals. Brown the tail; don't blacken it. This will take about 4 minutes for each side, depending on the heat of the coals and size of the tail. Slice the browned tail into 2-inch strips across its width. Season generously. Serve hot.

Serves 4 as a main course or 8 as a side dish.

The long-handled tool for cooking hamburgers can be found at camping equipment stores. As for finding a beaver tail, you are on your own.

SASSAFRAS
(*Sassafras albidum*)

Early colonists did not only eat hog meat as ham or bacon. If they were lucky, they might also eat fresh pork, seasoned with fresh local flavors such as sassafras. And for a short window of time, the leaf of the sassafras tree added flavor to savory dishes and also sold like luxury hotcakes back home in London.

The first British settlers in the 1600s eagerly sought valuable trade goods to ship back home to England. Indeed, profit motivated their adventurers, the optimistic investors who, rushing to replicate the Spanish gold and silver treasure from South America, underwrote the first expeditions of English colonists. Most settlers arrived hoping for opportunity, even the ones who emigrated primarily for religious reasons such as the Separatists (the ones we call Pilgrims) and the early Puritans in Massachusetts.

Tobacco, furs, fish, and lumber became vital exports within the first generation of colonization, with the noxious weed of Virginia leading the list in terms of value. But a medicinal herb and culinary flavoring would rank second only to tobacco as the most valuable export from the mainland colonies in the early years, and it's one that Columbus would have kicked himself for missing, if he could have lived another hundred years.

It was the dried leaf of the common sassafras tree, and at first Europeans hoped it was a miracle cure. Sassafras is not a native word and may derive from the plant name of *Saxifraga*. But the word makes a wonderful, swishing sound when spoken, which perhaps added to its magical reputation.

Algonquians along the East Coast had a variety of uses for sassafras. Village women harvested the mitten-shaped leaves in midsummer and dried them. Then the women crushed the brittle leaves into a powder that might be stored in decorated deerskin pouches. Used as a flavoring herb in traditional native dishes, sassafras also frequently appeared in a native healer's collection of medicines. Healers used sassafras as an aromatic tea for lung ailments and for fevers. They grated the root bark as a treatment for kidney disease and rheumatism.

When white settlers observed the indigenous use of sassafras as a healing agent, they thought they had stumbled upon a miracle cure, especially for a disease that had literally plagued the Renaissance era: syphilis. For a remarkable moment in the seventeenth century, Virginia sassafras was the second largest export from the British American colonies after tobacco.

Unfortunately, the combination of good marketing and magical thinking did not turn the leaves into effective medicine for the particularly virulent form of syphilis that grimly reaped more than one million Europeans during the sixteenth and seventeenth centuries. Sassafras quickly returned to its minor status as a local culinary herb once it proved no match for sexually transmitted disease. Later, Americans used it as a flavoring for temperance drinks such as sarsaparilla and root beer.

Contemporary Creole cooks in Louisiana still add sassafras powder to their gumbos and other dishes, only they call it filé. I add powdered sassafras, along with sage and thyme, to a simple roast chicken. In New York, small sassafras trees grow in the stony ground along the edge of our fence, and I can harvest a few leaves easily. And with a few keystrokes, you can order it online.

◀ BITE 16 ▶

❧ ENGLISH GARDEN ❧ HERBS AND VEGETABLES

I know, I know. This is supposed to be one Bite, or at least a single major ingredient, not a whole garden full of edibles. But in this case, I chose the place where a food group grew as the organizing principle for this section, rather than the food group itself, because the members of this group had (and still have) so many interrelated purposes for both food and health that it's impossible to separate them out individually.

And you, dear reader, should visualize the vegetables and herbs growing together, situated ideally on the south side of an early colonial house—where they would get the most sun—just a few steps from the door to the single

main room or, if the family was prosperous and well settled, to a separate kitchen.

The first colonists who emigrated here in the seventeenth century—from Spain, England, Holland, France, and Sweden—brought seeds and rootings across the Atlantic in copious amounts. Families nurtured apple and pear "slips" (baby trees), along with perennial herbs like rosemary, sage, thyme, and lavender.

They planted small orchards and garden plots, known as kitchen gardens, adjacent to their rough houses. In New England, larger orchards, pastures, and crop fields might be farther away. This is where men worked, although both men and women pitched in at harvest time. Women tended the vegetable and herbs,

A drawing of the sage plant.

perhaps with a hen house, in the kitchen gardens right next to their dwellings.

Maize and pumpkins grew out on the farm fields because they took up a lot of room. Most of the foodstuffs grown in the small gardens near a dwelling originated in Old World *potagers*, small kitchen gardens often symmetrically planted. Carrots, cabbage, green peas, leeks, spinach, beets, radishes, turnips, garlic, onions—these common vegetables joined New World beans and a wide variety of herbs in the garden bed.

Many of the common vegetables in these small plots kept well in cellars even after frost. People ate spinach and lettuce in season, before they "went to seed" or "bolted." But root vegetables and cabbage played an important role in the colonial diet because they kept well after the growing season was over.

WHO WERE THE REAL KITCHEN GARDENERS?

In the Puritan North, houses clustered together into villages to be near the church and each other. In the South, families built their homes next to their fields for closer access to the cash crops they cultivated and relied on for their livelihoods. But in both regions, it was generally women's work to take care of the vegetable garden.

When a household could afford the extra laborers, servants, both free and enslaved, performed the heavier work in the garden plot. Although white colonial families held slaves throughout the British colonies, enslaved African or American Indian captives were much more common in the South where the demands of plantation agriculture and the extended growing season made slave labor economically more rewarding. In fact, climate and labor systems were directly correlated. Slavery as a formalized system evolved during the 1600s, becoming more rigid and more vital to the South's plantation economy as the century progressed.

In New England, some young white servants might be related to their mistress or master, who had promised to provide training and guidance. Others were neighbors' adolescent children, who required extra training or needed to earn their own keep. In the colonial period, society understood that adolescents might take direction better from an adult who was not their own mother or father.

Others emigrated with indentures, contracts for seven to five years of work in exchange for payment for their ocean crossing. And sprinkled among these white servants were people of color, some free and some enslaved. But in most cases, a colonial housewife in the North tended her garden plot with very little help except for the older members of her typically large brood and maybe a neighbor's teenager.

Fat old carrots and mature parsnips might turn tough and bitter; still, they were edible in January, when stored in a well-built root cellar, and added vitamins and color to the scarce bland winter food available.

Vegetables played second fiddle to the meat and bread on the table. They could be served hot or cold, with a lot of butter or a little oil and vinegar, some chopped herbs, and plenty of salt, if salt was available. During the first generations of settlement, "sallet" was the name for a vegetable dish, hot or cold, whether you were in the colonies or in England at the time. Our word "salad" comes from this term, which simply means "salted."

Other times, housewives added vegetables to stews, savory pies, and soups, just like we add them today. Rarely were vegetables served raw, although they might be served cooked and then cooled, or pickled with apple cider vinegar. In general, the first English settlers considered raw vegetables an unhealthy if not savage alternative.

Many of the first settlers arrived with little agricultural experience. They followed traditional practices that varied according to the regions from which they had originated. On top of these traditions, the realities of life in the New World forced their own requirements, such as hoeing rows around stumps in the newly cleared land. Not an easy feat for anyone, much less someone new to farming.

Colonists grew herbs that they brought over from England, among their vegetables for culinary, medicinal, and "strewing" purposes. Strewing herbs—such as santolina, rue, lavender, and pennyroyal—smelled clean and helped keep the houses and clothing chests smelling fresh as well. They also kept insect pests at bay to some extent. The name "lavender" reflects the Latin verb "to wash," and sprigs of lavender as well as sachets were folded into fresh linen after it had been cleaned.

Medicinal herbs, such as chamomile for soothing upset stomachs or sage for headaches, served a dual purpose in the kitchen, where sage joined traditional culinary herbs—parsley, rosemary, thyme, chives, and dill—in flavoring primarily savory dishes, while chamomile made a mild all-purpose tea. Some herbs helped with moods and character issues. Apothecaries in seventeenth-century England, for example, recommended borage tea for people who were nervous or timid, because it gave them extra courage. (Perhaps because it rhymed!) Herbal manuals also prescribed the use of herbs whose leaf or root resembled the afflicted body part to heal wounds, injuries, or ailments. This is known as the doctrine of signatures. So for example, doctors might recommend a decoction of lungwort with its lung-shaped leaves for chest complaints.

An important feature of the goodwife's garden, herbs grew side by side with vegetables, not in a separate space. They played an integral role in food and flavors as well as household management. Women harvested the leaves from most herbs just before they flowered, when their scent or taste was strongest. Then they would hang the herbs in bunches to dry, ideally in small muslin bags from the ceiling beams in the upper loft. Once dry, the leaves could be ground into powder and stored in small stoneware containers or left, crumbled, in the muslin bags, much like large versions of the sachets people still use today to perfume their drawers and homes.

Today, herbs retain their appeal, and people still grow them, buy them, and use them in droves. I abandoned my small vegetable garden years ago, but I still grow herbs. Now they thrive among my flowers instead. I love their textures, tastes, and scents. And homegrown herbs freshen the flavor of any store-bought food, making it taste 100 percent homemade.

During the reign of Queen Elizabeth I, a young Englishman named Fynes Moryson mentioned cockaleekie soup in his multivolume travelogue about the New World.[5] Recipes for nourishing, thick soups like this one traveled across the Atlantic with the first settlers, warming families with traditional English ingredients and memories of home. The inclusion of minced prunes reminds us that the medieval fashion of mixing fruits and meats in main dishes lasted well into the 1700s and still lingers in heritage holiday foods like mince pies.

Following is an updated version of a very old recipe that relies on garden veg-etables and herbs. It does not require any fancy technique, but it does take time. Of course, people often weren't in a big hurry back then. To say that the pace of life was slower in the colonial period is an understatement—I don't think modern people can grasp just how quiet, slow, dirty, and dark (especially in the winter) the 1600s were in this country, although punctuated by merry feasts and significant alcohol consumption.

COCKALEEKIE SOUP

Scotland claims cockaleekie soup as a homeland dish, although a chicken and leek stew could easily be associated with Wales, where leeks serve as a regional symbol. Traditionally, pearl barley would thicken the broth, but you may prefer rice. Incidentally, if you opt for the latter, white rice would be more historically accurate than healthier, modern brown rice. Go ahead and improvise—add brown rice if you want. That would be the authentic seventeenth-century attitude.

INGREDIENTS

- 1 2- to 3-pound chicken
- 8 cups chicken stock
- 1 cup chopped carrots
- 1 cup white wine
- ½ cup minced celery
- 2 teaspoons allspice
- 2 teaspoons dried thyme
- 1 bay leaf
- 1½ cups pearl barley or white rice
- 4 leeks
- 1 cup diced, pitted prunes
- ½ cup fresh parsley, minced
- Salt and freshly ground black pepper to taste
- 1 cup heavy cream

DIRECTIONS

Place the chicken, chicken stock, carrots, white wine, celery, allspice, thyme, and bay leaf in a Dutch oven or heavy pot. Bring to a boil over medium-high heat, then reduce to simmer, cover, and cook over low heat for 50 minutes. Remove the chicken carcass and let it cool.

Stir the pearl barley or rice into the broth, cover, and simmer for 30 minutes. Meanwhile, rinse the leeks well under cool water. Trim tops and ends and remove outside leaves, then chop up the remaining 4 inches of the midsection. Add the leeks, prunes, and parsley to the mixture, and simmer covered for about 20 minutes, keeping the heat low. Make sure the barley is thoroughly cooked and soft. Add salt and pepper.

Cut the meat off the chicken carcass, discarding the skin, and chop meat into bite-sized pieces. Stir chicken pieces into the soup, and heat thoroughly over low heat. Adjust seasonings. Serve with a splash of heavy cream in heated bowls, and with brown bread and butter alongside.

Serves 8 as a main dish.

Rather than using a whole chicken, you can substitute 2 boneless, skinless chicken breasts and 4 thighs. Chop them up, and cook them in the pot if the whole chicken carcass routine is overwhelming. You'll miss some rich flavor and authenticity, but that's up to you. Also, you can stop before adding the meat off the chicken carcass and refrigerate the soup overnight or for up to two days. To serve, reheat the soup over low heat, adding more chicken stock if it is too thick. Then continue to add the chicken and adjust the spices. The soup freezes well and makes a nice gift for New Year's Day.

BITE 17
❧ COW'S BUTTER ❧

Dairy cows and the milk, butter, and cheese that they produce were an important part of the English diet, so it's no surprise that colonists brought cows across the Atlantic to guarantee fresh dairy products. People and cows shared a long history in Europe and Asia as well. Humans domesticated cattle around eight thousand years ago, and cow's butter followed soon after. Cows figured as important symbols in ancient Greek and Roman mythology, but it was in cooler northern Europe that dairy culture played a particularly important economic role. Scandinavia exported butter as early as the twelfth century!

In the New World, Columbus introduced the first cows in 1493 on his second voyage, and subsequent Spanish explorers and conquistadors brought cattle too. Soon the Indians began to raise cattle for meat and for dairy, with fresh cheese—*queso fresco*, as it is known in Mexico—becoming a frequent addition to traditional corn and beans.

Dairy played an important part in northern European foodways, and

that role crossed the Atlantic with the cattle that traveled with the colonists. In Virginia's English settlements, newly introduced cattle foraged happily.

Each owner notched the cows' ears in a personal pattern, so the ear marks served like a brand to tell the different owners' cows apart. Once the inexpert Pilgrims figured out how to provide for cattle over the long New England winters, the herds grew there as well.

Dairy cows in seventeenth-century Virginia and New England were smaller than the long-legged commercial milking breeds today. The Red Devon breed, popular in the early colonial period, were multiuse animals, reared for milking and meat, and also trained as oxen. Their small size made them easier to handle and perfect for a household that could only afford one or two cows. They gave rich milk with a very high butterfat content, even more than a Jersey or Kerry cow today.

Women and girls generally were in charge of a household's milch cow, hence the proverbial dairy maid or milkmaid in old English rhymes and folk tales. (Of course, eventually farms developed that specialized in milk production, where the entire family and hired hands focused on dairy products.) In the spring and summer, when plenty of green grass grew, the cows gave the most milk. But that was also the warmest time of the year, when milk quickly soured.

The rich milk from a Devon cow that's left over after skimming is a far cry from the fat-free milk we call skim today. When I worked at Plimoth Plantation, a living history museum on Cape Cod, I occasionally drank the so-called skimmed milk, much richer even than what we call whole milk, produced by the Devon cow there. The milk from this heritage breed contains much more butterfat than the milk from a black and white Holstein, the type that commonly populates modern dairy herds in America.

How to preserve the extra milk without refrigeration? That is when salted butter and cheese played a starring role as key methods of preserving fresh milk. Settlers knew enough to keep the dairy utensils scrupulously clean—a tough job in the seventeenth century, but of crucial importance to prevent milk from going bad. After milking the cow, a dairy maid would pour the fresh milk into wide bowls, cover them with a clean cloth to keep the flies away, and let them sit until the cream rose to the top. Then she would skim the cream off and place it in her butter churn. The remaining liquid in the bowls was skimmed milk, the origin of our term "skim milk" today.

The dairy maid would sit on a stool, and stand the wooden butter churn filled with cream on the floor, steadying it between her knees. She would pump the dasher up and down, until the butterfat broke away in clumps from the buttermilk. Then she would remove the dasher, scrape the chunks of butter off it, and wash the butter in cool water until the water ran clear and all the whey was rinsed away.

In the summer, the butter was golden and appetizing looking, but in the fall, once the cow was no longer eating fresh grass, the butter was white, like lard. So some households dyed the butter to make it more attractive. Many of us today might be surprised that people in colonial America dyed their food. After all, our own prevalent practice of adding dyes to our foods is coming under serious scrutiny.

But the golden color of butter in the spring and summer was highly desirable. So even modest families might dye their winter butter with dried flower petals. Along with vegetables and herbs, a family might grow calendulas, or "pot marigolds." After harvesting and drying the bright gold petals, a daughter could rub the petals into a yellow powder. This powder could then be mixed into the white butter, giving it a golden color.

Once the dairy maid finished the butter, she packed it in small earthen pots, layering it with salt to preserve it. Stored in a cool cellar or a springhouse, the well-salted butter could last several weeks. The leftover buttermilk was used for cooking or as a beverage.

Dairy farmers also produced homemade cheese. Artisans from Britain and continental Europe brought their skills with them when they emigrated. Cheese makers sold big wheels of cheddar in New England, and the Dutch settlers in New Netherland (later called New York) made balls of a variation on traditional Gouda. Perhaps the cheeses weren't exactly like the stuff back home, but they were tasty, nutritious, and helped preserve milk's goodness for the lean winter ahead.

Butter would remain a key component in American cooking, as an ingredient as well as a fat for frying. Early cookbooks contain recipes for everything from cakes to oyster sauce that specified butter by the pound. Its use far surpassed that of olive oil and the other vegetable oils we often use today for two obvious reasons. First, those oils, especially olive oil, simply weren't nearly as available back then, if at all, since they had to be imported. And secondly, the colonists loved butter.

Of course, butter wasn't the only fat popular in the past. Beef suet and pork lard have been featured in baking and deep-fat frying throughout American history as well.

For North Americans—American Indians, Europeans, and African Americans—preserving and eating fat continued as a food priority into the twentieth century. People were active and labored hard. They burned many calories just getting through the day, and eating foods and dishes high in fat helped them keep from getting too thin or weak. And while food supplies grew plentiful, especially for colonists along the Eastern Seaboard, an unusually lengthy winter or an extended drought locally could drastically impact the variety and amount of food available on the family table. So butter and other fats helped people survive particularly harsh winters.

Even in a land of food abundance for most, many of us are still genetically coded to crave the rich taste and sumptuous mouth feel of flavor-delivering fat. Personally, I try to eat very little butter, but when I do, I like the kind with the really high butterfat, because otherwise what's the point? And while butter may border on the sinful for many people today, it's important to remember that only recently has it been seen as an evil temptation. A golden, rich slab of salted, extra-high-fat butter was a heavenly sight in the early days of our colonial past.

BITE 18
★ EEL ★

For centuries, western Europeans adored eating eels—fresh, smoked, or salted and preserved. England's King Henry I legendarily died from eating a "surfeit of lampreys," a species of eel and a delicacy he clearly enjoyed to its utmost. Cockney street hawkers sold jellied elvers, or baby eels, fished from the Thames estuary, and Londoners can still dig into a plate of this local dish in Cheapside. In Spain today, some traditional tapas features a dish of tiny, whole *angulas*, drenched in hot olive oil pungent with garlic, truly delicious with a slice of crusty bread and a small glass of red wine.

Eels are not as fashionable in the United States, for no good reason that I can understand except that Americans seem to associate eels with snakes and wrinkle their noses in disgust at the very thought. That said, a few restaurants serve eel proudly, and our national cuisine is better for it. Having dined on delicious conger eel in Chile, I am convinced that it is frequently served in U.S. restaurants masquerading as high-priced sea bass. Perhaps the rising fashion for Japanese sushi, including *unagi*, freshwater eel, may be positively shifting a small percentage of people's views in favor of this tasty seafood.

Like their cousins across the sea, early British Americans considered eels a gourmet's dream. American Indians from the Patuxent tribe taught the settlers how to harvest eels from the muddy inlets along Cape Cod with their bare feet. Settlers also discovered local streams roiling with young eels. In Massachusetts, the colonists baited woven-rush fish traps, known as weirs, placed them in the stream, and soon brought up a wriggling basketful. Yum!

What I found particularly stunning about this process is the settlers' choice of bait: lobster. Today, I would hazard a guess that most Americans would prefer to bait a lobster trap with eel rather than vice versa. But in the early years of British settlement, lobsters were abundant and enormous. Colonists harvested giant lobsters with spreads of five feet from claw to claw, right in Plymouth Harbor. Apparently these mammoth crustaceans tasted bitter, with unappetizingly tough flesh. But they were the perfect bait for the delicate eels so coveted by the early settlers and praised by kings and Cockneys back home in England.

Modern Americans may be slow to embrace eels as a dining pleasure, but the Asian market is booming. Licensed fishermen in Ellsworth, Maine,

sold elvers at almost $2,000 a pound to Chinese buyers in the early spring of 2012, when the little baby eels began their run to the sea.

OLD EEL PIE

Englishman John Murrell published a cookbook in 1615 that included this recipe for eel pie. While we don't know for sure that any colonists owned a copy of A New Booke for Cookerie, *food historians believe that housewives and cooks might have used this type of recipe or "receipt."*

TO BAKE EELES.

Cut your Eeles about the length of your finger: season them with Pepper, Salt, and Ginger, and so put them into a Coffin, with a good piece of sweet Butter. Put into your Pye great Razins of the Sunne, and an Onyon minst small, and so close it and bake it.

(Excerpted from John Murrell: A New Booke of Cookerie. *London: London Cookerie, 1615.)*

"Coffin" simply means box here; it does not refer to a place for a corpse, so don't worry. (Though in a sick sense, it is a coffin for the eels.) In the 1600s, a coffin was a deep pie shell, often made of whole-wheat flour pastry, which could be a little tough. The pastry would line a 3- or 4-inch deep pan that could be covered and set among the coals to bake. The coffin was more important for cooking than for eating, unless you were an Old World beggar. In that case, you were lucky to get the coffin crust.

BITE 19

❧ PERRY ❧
(Pear Cider)

The New England settlers were delighted to find an abundant supply of eels, but actively complained about the lack of beer. Many scholars have delved into the jolly topic of beer, one of the oldest documented foods. In fact, one of the earliest known recipes of any kind is for beer, discovered by archaeologists studying ancient Sumeria in the Middle East. In England during the 1600s, everyone drank beer, young and old alike, with variations in alcohol content, flavor, color, and viscosity. Beer was usually healthier than drinking well water, which often was contaminated.

When the English settled in the New World, they were suspicious of the freshwater resources and longed for the good, honest beer they had left behind across the Atlantic.[6] Eventually, they would grow the required ingredients and actively recruit experienced brewers to join their settlements. Brewers and alewives produced adequate beer for their neighbors, but the process was more hit or miss than in Britain. The problem was that wild yeasts occurred naturally in America and could ruin the fermentation process. Also, summers in the colonies were hotter than in Britain, and traditional beer would sour more easily. When the German immigrants arrived in the nineteenth century with a more reliable recipe, brewing in America really took off.

But until that watershed moment in American history, colonists frequently preferred hard cider and perry to local beer. Within a few years of their arrival, settlers in the North and South had planted apple and pear orchards and quickly started harvesting those fruits for more than their nutritious food content. The production of apple or pear cider, known as "perry" or "peary," did not require the same finesse as beer brewing, and cool fall weather kept the fruit juice from turning bad, allowing the natural fermentation process to take place. So it's easy to see why this hard cider became such a popular drink.

Like their contemporaries in England, the settlers consumed a lot of alcoholic beverages, whether they were New England Puritans or southern planters. Early records document the importance of wine, Madeira, and brandy, but only the wealthier citizens regularly served these pricey, imported drinks at their tables. The average colonist generally drank hard cider or perry.

Today, you easily can find pear brandy or schnapps at a local liquor store, and hard apple cider, often imported, as well. Perry, however, requires a bit of a search. I thought to give hard pear cider the proper respect here because it was once so ubiquitous, almost like certain brands of soft drinks today, yet most Americans have no idea what it is (or was).

Unlike today's big soft-drink brands, perry was, of course, organic and made by hand, fermented from the juice of the crisp, hard pears of late fall, not the sweet, slurpy Bartletts that are so delicious in late summer. Today, Blackthorn, the English hard cider company, offers a perry in this country, and there are local sources online as well. Sip some cool, dry perry and you'll encounter a taste long forgotten in our country, lingering in the rustic past. I hope it makes a comeback. Make sure you buy real perry, not hard apple cider flavored with pear juice. Sadly, the American retailers are more apt to use this technique, while the British sell the real stuff.

BITE 20

TURKEY

(Meleagris gallopavo silvestris, M.g. intermedia, M.g. Mexicana)

Of all the foods associated with the first British colonies in America, perhaps the turkey wins the prize for most iconic. Ironically, while turkeys are indigenous to North America, some settlers may have encountered a domestic breed of turkey in England before departure. The Spanish explorers brought turkeys back to Europe more than a hundred years before the date of the legendary first Thanksgiving in Massachusetts.

A large showy bird, the turkey easily adapted to life in the European barnyard and served as a plattered centerpiece at feasts. Wealthy Renaissance hosts, like their predecessors in ancient Rome, loved to impress their guests with big roasted fowl, such as swans or cranes. Unfortunately, these elegant birds were not too tasty, although they made a great-looking presentation. The turkey's arrival solved this issue. It looked just as glorious on a plate as a roasted swan, was not expensive to raise, and provided flavorful meat with rich drippings—always a plus.

When this big bird first showed up in London's markets in the mid-1500s, people wondered what is was, and where it came from. It certainly seemed foreign. Contemporary documents tell us that merchants from the Levant—Turkey—first brought turkeys to Britain. The Turkish merchants had purchased these big birds originally from Spain. The English named this poultry "turkey" because they believed that was where these birds originated.

It was a good marketing technique as well, for Turkey represented the exotic East in the imagination of the British in the 1500s, and given human nature, exoticism often is interchangeable with desirability. The English easily could have called it "India bird," which is roughly the translation of its French name and its Turkish name as well. The English named the guinea fowl, a smaller bird, correctly after its land of origin, the country Guinea in West Africa. Guinea hens had been imported by the Portuguese traders. If history was always predictable, we'd call these speckled fowl "portugals" today.[7]

In America, wild turkeys ranged from Central America all the way to eastern Canada. Long ago, they roamed the forests and clearings of the Spanish and British colonists, their brown mottled plumage camouflaging them from their predators. First domesticated by the Aztecs, the big bird originated in Mexico. Indians of the Southwest roasted it on a large spit and used the feathers decoratively in clothing. The smaller Osceola turkey, later named after the charismatic Seminole leader, inhabited only the Florida peninsula. Wild turkeys were so popular a food source that they became over-hunted and rare by the end of the 1800s.

To understand the importance of turkey in our culinary heritage today, let's take a look at how it became the iconic food of America through its association with Thanksgiving, a vivid part of our shared, almost mythic past. The idea of Thanksgiving is based in part upon the natural inclination of agrarian groups of people to hold a festival in thanks for the harvest, and we humans have been celebrating the gathering-in of crops for millennia.

> *Today, wild turkeys are making a comeback, slowly in some places and with a vengeance in others. I often see small flocks by the side of the road on Long Island, diligently pecking in the high grass.*

The term "Thanksgiving" originally included serious religious dedication, with several hours spent in church—and it started long before the famed feast between American Indians and colonists in Massachusetts. In 1519, at St. Augustine, Florida, the Spanish celebrated Thanksgiving with pork and chickpeas brought from the Old World. According to contemporary sources, a harvest dinner shared by Spanish settlers, missionaries, and American Indians took place in Santa Fe, New Mexico, in the 1590s. At the Berkeley Hundred settlement in Virginia in 1618, the English dined on ham and gave thanks for their safety and survival.

Of course, the 1621 harvest feast in Plymouth, Massachusetts—where the "Pilgrims" (the term is in quotes because they wouldn't have labeled themselves that) were joined by ninety Wampanoag warriors—is the big dinner remembered every November. We know that four Englishmen went out hunting for that celebration and brought back unspecified fowl, which could be anything with wings—duck, geese, partridge, or yes, even turkey.

They also may have served eel and shellfish, plus foods based on the Three Sisters, which their indigenous neighbors had taught them to grow with such success. We know that the Indians brought venison. Cranberry sauce, which requires so much sugar, would not have made an appearance, although stewed pumpkin sweetened with honey or maple syrup may have been shared.

The historical record about the harvest feast we celebrate as the first Thanksgiving does not specify a turkey, but it is clear that there was plenty of food at the celebration. As the colonial period progressed, the tradition of a harvest festival continued, particularly in the Northeast, where it was observed at different times in different colonies.

Families traveled to be together and dined on turkeys as well as chicken pie, ham, and game. Women worked hard for weeks to present a table laden with different dishes, along with an abundant array of fruit pies and cakes, and distributed gifts of food to the poor. After the American Revolution, young families emigrated from New England, looking for farms of their own in the western territories, and brought the Thanksgiving tradition along with them.[8]

This particularly home-centered holiday grew in cultural importance before the Civil War, when it was championed by Sarah Josepha Hale, the Martha Stewart of her day. Cookbook author, novelist, and magazine editor, she published recipes for roast turkey with stuffing and pumpkin pie, along with editorials favoring the creation of a new national holiday. At that time, governors of several states declared Thanksgiving at some point in the late fall.

Hale encouraged President Lincoln to make the feast day a single national holiday for all, uniting every American. With his uncanny political timing, Lincoln authorized this quintessential American celebration in 1863, just as things were looking up for the Union in the Civil War. (We'll learn more about this in Bite 44 on Thanksgiving.)

One reason why our images of Thanksgiving reflect the Pilgrim legend is that New England (and the North more generally) culturally predominated in the United States in the years after the end of the Civil War in 1865. And this was the era when popular artists created the images of our mythic New England forefathers and foremothers gathered around a scenic table, complete with a big turkey roasted to a golden fare-thee-well. So it is the story of the Wampanoags and the settlers of Plymouth, not Jamestown, Virginia, and certainly not St. Augustine, Florida, that schoolchildren have reenacted for more than a century. Quite possibly, other similar Thanksgiving celebrations between European settlers and American Indians occurred as well. We just didn't hear about them.

Most of the turkey available commercially today descends from a breed developed in the 1930s, the Broad-Breasted White. Farmers colloquially called this all-white bird, with extra breast meat, the "Mae West" turkey, after the buxom blond movie star popular back then. These poor things can't even breed naturally anymore and require artificial insemination to reproduce.

I have found that the heritage-breed turkeys available online provide an easy introduction to early American authentic food. The Narragansett is the oldest domestic breed, and very tasty, but the Bourbon Red or the Bronze are also good. Because they have denser skeletons, remember to buy a heavier bird than you would if you bought the supermarket kind. Don't be surprised by their mottled skin color, which reflects the color of their feathers. Their skin won't glisten a whitish pink, like the genetically enhanced Mae West turkeys you see lined up at the supermarket. Just one caveat: the Broad-Breasted White is mass-produced and much cheaper to buy. The heritage turkeys come from family farms and are priced significantly higher.

Whether you choose the old-fashioned heritage bird or the twentieth-century Butterball style for your Thanksgiving dinner, you are reenacting a festive tradition that is four hundred years old. Enjoy participating in a uniquely American meal, laden with meaning and myth.

chapter 3

BREAKFAST WITH RUM AND TEA

From Colonies to Country

hanges in food and domestic life moved slowly in colonial America. Patterns of rustic simplicity and limited food sources repeated themselves as families pressed westward on the ever-shifting frontier toward the Appalachian Mountains. By the mid-1700s, the comfortable lifestyles of town dwellers along the Eastern Seaboard approximated that of their British cousins, and these Americans began to clamor for imported goods and ingredients to enhance the rustic resources available.

But the British Empire's need for more revenue drove new taxes up and imposed limitations on the very items the colonists longed for the most, such as sugar and tea. This frustration proved to be an important factor in the fight for American political independence. Our culinary independence had already begun with that staple called maize.

BITE 21

❧ CORN AGAIN! ☙

You may have thought we were finished with this grain already, but we're just beginning. And it's worth revisiting this vital food here to see how ubiquitous it became in colonial cuisine. We know that maize did not impress the first settlers, but once they adapted to life in the New World, most of them adopted "Indian corn" with gusto. Adding cornmeal to expensive wheat flour reduced the cost of baking traditional wheat bread recipes. It provided the basis of delicious Indian pudding, handy johnnycake (or journey cake), and indispensable cornmeal mush.

Up and down the colonies, British Americans ate corn. When they traveled locally or long distances to visit family or to conduct business, they packed johnnycakes for the road. Cooks used yeast or beaten eggs to raise the round, thin cornbread—since reliable baking powder did not exist—but often the recipe was a very simple one. Wrapped in oiled cloth or parchment, johnnycakes could last a few days without going stale. If it did harden, the cornbread could be crumbled and then soaked in milk or cider, creating a sort of porridge.

Everyone had a favorite corn-based dish. Hot cornmeal porridge baked with milk and molasses, known as Indian pudding, could be served at any meal. Thrifty farmwives served cornmeal mush at the start of the midday dinner, the main meal of the day, to fill the empty edges of hungry stomachs, similar to the way bread is put on the table before a meal today. After family members and laborers finished their mush (or loblolly, as it was colloquially called), a housewife would spoon out the main-course stew.

When guests were entertained in the homes of the gentry, servants might place a warm bowl of Indian pudding on the table as part of the second course at dinner. In eighteenth-century Virginia, a colony known for its generous hospitality, the dish would join an array of sweet baked goods and other puddings set out in front of the seated guests. Indeed, every course featured one or two items based on maize. In colonial America, cooks presented corn, cornmeal, or corn flour at breakfast, lunch, and dinner.

INDIAN PUDDING

Hot cornmeal mush baked with rich milk and molasses has pleased crowds of hungry Americans for centuries. The name "Indian pudding" reflects the main ingredient in Indian corn; the milk and molasses were European imports. You can still find it on menus of New England restaurants that pride themselves on heritage cooking. Champney's, the restaurant and tavern at the Deerfield Inn in Deerfield, Massachusetts, still serves an excellent Indian pudding.

BREAKFAST HOE CAKES

George Washington, a native Virginian, enjoyed French wine and imported brandy, but he was a born-and-bred British American who loved cornmeal pancakes. Called "hoe cakes," they were his favorite breakfast, especially when served swimming in butter and honey. These traditional flat cakes were fried on a flat griddle or the iron plate from a large hoe, hence the name hoe cakes.

Historic cooking expert Nancy Carter Crump provides an authentic and delicious recipe for hoe cakes in Dining with the Washingtons.*[1] It is also quite time-consuming for today's cook. My modern version is easier and quick to make, relying on baking powder instead of yeast to make the batter rise. This is essentially your own hoe cake mix, and it keeps in a sealed container or zippered plastic bag for a long time. I hope George and Martha Washington would approve.[2]*

INGREDIENTS

► 1½ cups white cornmeal

► ½ cup unbleached white flour

► 1 tablespoon baking powder

► 1 teaspoon baking soda

► 1 teaspoon salt

► 2 eggs

► 1⅓ cups whole milk

► 2 tablespoons canola oil or melted butter

► Butter, if desired, and honey, molasses, or maple syrup for serving

DIRECTIONS

Stir the cornmeal, flour, baking powder, baking soda, and salt in a large bowl. Beat the eggs in a separate bowl, and stir in the milk and oil. Add the liquid ingredients to the dry ingredients, and stir well. Add a little more milk if the batter is too thick, thinning it to the consistency of any pancake batter.

Preheat oven to 200°F, and have a nice platter or large plate ready so you can keep the first hoe cakes warm while you prepare the rest of the batter.

Drop batter by scant ¼ cup (I use a soup ladle) onto a lightly greased hot griddle. Turn when bubbles form on the top of each hoe cake. Cook until medium brown on both sides. Add each finished hoe cake to the platter, and keep warm in the low oven. I like to warm our individual plates at the same time, especially in the winter. Serve on warm plates with butter and honey, molasses, or maple syrup.

Serves 6.

Alternatively, you can visit Mount Vernon on February 22, when the knowledgeable staff serves fresh hoe cakes in honor of our first president's birthday.

BITE 22

❧ DOUGHNUTS, ☙ WAFFLES, AND COOKIES

You may think we have a giant national sweet tooth today, but the early American settlers were no less fond of their sweet treats. The Dutch contributed to a national love of sugary baked goods. The English and the Spanish weren't the only Europeans to colonize North America early on. The French sent explorers and missionaries, whose names dot the maps of modern Canada and the Unites States, particularly in the Great Lakes region and the Mississippi River valley. The Dutch founded New Netherland right in the middle of the Atlantic coast and up the great Hudson River. Their harbor town, New Amsterdam, would eventually become the vibrant powerhouse called New York City.

Although the English took possession of New Netherland in 1664 and renamed the colony New York, many Dutch settlers remained. Farm families in the early 1800s still spoke Dutch. Several common American words, like "boss" and "stoop," are Dutch in origin. Their influence lingers in place names like Brooklyn and Kinderhook, and in family names like Roosevelt and Vanderbilt.

The Dutch also contributed three all-American foods: doughnuts, waffles, and cookies. It would be inaccurate to say they invented these sweet treats. Honey-drenched delicacies were featured in the feasts of ancient Babylonia. English cooks baked sweet biscuits

A waffle iron.

that were tasty. But the Dutch produced rich cookies with almonds and spices, cocoa and walnuts, molasses and raisins. They dominated the international spice trade, and their ingredients reflected this access to the exotic. They decorated cookies at Christmastime, pressing stiff dough into sharply carved molds with images of windmills, crowns, and crosses. If you visit Holland today, you will still find these types of confections everywhere.

Dutch settlers brought round waffle irons with them across the Atlantic, with long handles for holding over a fire. Their British neighbors enjoyed this delicious food too. Colonial Williamsburg has waffle irons from the 1700s in their collection. But the Dutch were the waffle masters, serving them with savory dishes, such as creamed chicken and onions, as well as with fruit and sweet buttery toppings.[3]

They also made doughnuts. In the colonial period, doughnuts were chubby circles, not the inflated rings we're used to seeing today. A baker would drop a ball of stiff batter or dough about the size of a large walnut into hot pork lard and fry it brown. They called these deep-fat-fried calorie bombs *oleykoeks*, or oil cakes, serving them with a dusting of confectioners' sugar and cinnamon. British Americans adopted the recipe, but called the treat a "dough-nut."

An American seaman cook, Hanson Gregory, claimed to have created the first doughnut cooked intentionally with a hole in 1847, more than two hundred years after the Dutch arrived here. He said the hole allowed the dough to cook more evenly in the hot lard. Today, Americans consume doughnuts with enormous relish, chowing down on over ten billion a year.[4]

> *The director general of New Netherland, Peter Minuit, purchased a long, thin island called "Manahatta" at the mouth of the Hudson River from the Lenape Indians in 1626, christened it New Amsterdam, and designated it the capital of the fledgling Dutch trading post. And you've probably deduced correctly that that's where the name of the island of Manhattan came from.*

BITE 23

Wheat Flour

When we say "flour" today, we generally assume we're talking about some sort of wheat flour unless we specify otherwise. Wheat flour—from dark whole grain to bleached white—provides the basis for most baking in the Western world. The settlers from the British Isles and Europe brought this preference for wheat with them when they came to these shores. Even though colonists ate a lot of corn bread, wheat bread was the more desirable staff of life.

By the 1700s, several colonies produced enough wheat for it to become an important export. William Penn's tolerant Quaker region in what is now Pennsylvania attracted Germans and other groups like the Moravians who settled west of Philadelphia, seeking land and religious freedom. There, wheat thrived in the rich, sweet soil. In the colonial period, the farmers sent barrels of wheat flour to market on riverboats or oxcarts, headed for the city and points south to supply plantation-based colonies.

People of European descent craved wheaten bread, the whiter the better. White flour, a more expensive and thus more coveted product, might be saved for holiday baking or special meals. Whole wheat flour was cheaper, heartier, and more rustic. It also might include bits of husk. Back then, dietary advisers did not promote whole grains, but instead urged mothers to serve sickly family members white gruel.

As the eighteenth century progressed, dense porridges and blood puddings thickened with oats were falling out of favor in Britain. Wealthy, fashion-conscious families appreciated lighter cakes and pastries, perhaps influenced by French chefs who had starring roles in court kitchens. Colonists read about British trends in newspapers and heard about the latest foods from friends and relatives living abroad, and even from acquaintances at the local tavern. They bought English cookbooks that specified wheat "flower" far more frequently than any other finely ground grain. Soon it became the standard for British Americans and their colonial brethren, neighbors from other European countries. Finely ground wheat remains the dominant flour in America's kitchens today.

BITE 24
★ OXTAIL STEW ★

Some aspects of daily life in early America—serving waffles, for example, or ham—have stayed pretty much the same over the course of four centuries. Other aspects have changed radically, and transportation is one of the categories that has changed the most. Although you have to go to a living history museum to see one today, the oxcart of Biblical times was common in the colonies.

Oxen played a vital economic role in colonial America, all along the Eastern Seaboard and in Spanish North America too. An ox is simply a cow or a steer (castrated bull) trained to wear a yoke and pull a plow, a cart, or

Oxen were invaluable help on the farm, and their tails made an invaluable stew, especially for those with little to no means.

another heavy weight. Introduced to the American continent from the Old World, oxen dragged lumber, hauled hogsheads of tobacco, and helped raise beams in houses.

It was not horsepower but ox power that built the colonies. Yes, big draft horses had enormous strength—but cost a pretty penny. They ate twice as much, and they cost twice as much. Eventually, mules would become particularly important in the 1800s on long pack trips in the western territories. But the slow, stodgy, unromantic ox hauled the colonial economy from struggling settlements to thriving communities.

THE OX VERSUS THE HORSE

For ten thousand years, oxen had served as the workhorses on farms, in seaports, and along bumpy quagmires known as roadways all over the globe. It was no different in colonial America, where oxen provided the strength and stamina to clear fields and haul heavy loads. People bragged about their horses back then, the way they might today in equestrian circles. They had portraits painted of their beloved equines, and they loved horse racing too. The first horse track was built in 1665 on Long Island, New York.

Farmers may have pitted their oxen against other teams, in a pre-industrial version of a truck pull, but horse racing attracted the big crowds. People admired and made note of accomplished equestrians of both sexes as well. One of George Washington's personal strengths was his skill on horseback. Few (if any) ox-team drivers, no matter how talented, ended up in our history books.

For obvious reasons, the more muscular, the better when it came to oxen. Eventually, however, its hauling days over, the slaughtered ox would head for the stew pot. Tough and strongly flavored, the meat challenged the impatient or inexperienced cook. But the tail, with vertebrae surrounded by meat, could be simmered into submission by a knowledgeable farmer's wife in New England or a skilled enslaved cook down south—or in New York, for that matter. Even in New Spain (Mexico and parts of the American

Southwest today), oxen worked as vital draft animals as well. Today you will still find oxtail soup on menus in restaurants, but this working-class dish is now considered a rustic delicacy.

OXTAIL STEW

This adaptation of a traditional oxtail stew is similar to what would have been served at a home or tavern in the British American colonies. Spanish settlers in what is today's American Southwest ate oxtail stew as well. To make this a historically authentic dish, leave out the celery and the white potato. The Spanish colonists used more garlic than the northern Europeans and added red pepper, so you can choose the variation you prefer and present your creation in all its deliciousness as English Oxtail Stew or Spanish Oxtail Stew. Corn bread is a nice offering alongside either version.

INGREDIENTS

- Salt and freshly ground black pepper to taste
- 3 pounds oxtails with separated joints, bone in
- 2 tablespoons butter or olive oil
- 1 medium onion, chopped
- ½ cup chopped celery
- 3 cups beef stock
- 2 cups red wine
- 1 bay leaf
- 1 teaspoon dried thyme, or 1 tablespoon fresh thyme
- 1½ cups peeled and chopped carrots
- 1 cup peeled and chopped parsnips
- 2 medium sweet potatoes, peeled and chopped
- 1 medium white potato, peeled and chopped
- ½ cup minced parsley
- 2 teaspoons dried sage

DIRECTIONS

Salt and pepper the oxtail pieces generously. Brown them in the butter or oil in a heavy 6-quart Dutch oven over medium-high heat for 2 minutes on each side. Do this in batches so all pieces have a nicely browned exterior. Remove meat from the pot and set aside. Reduce heat to low, and add the onion and celery to the pot, stirring occasionally until translucent, about 3 minutes. Add the stock, wine, bay leaf, thyme, and oxtail segments. Bring to a low boil, cover, and simmer 2 to 3 hours over very low heat. Allow the pot to cool, and place it in the refrigerator or a cold shed overnight.

Remove the hardened fat from the surface of the stew the next day. Return the Dutch oven to the stove top. Add the carrots, parsnips, potatoes, parsley, sage, salt, and pepper. Bring to a low boil, then reduce the heat to simmer, and cover and cook for 1 to 2 hours, or until the meat and vegetables are very tender. Serve in soup plates, leaving the bone in each oxtail segment.

Serves 6 to 8 as a main dish.

For the Spanish version, add 5 minced garlic cloves with the bay leaf and thyme, and then 2 teaspoons chipotle powder or red pepper flakes with the carrots, parsnips, and potatoes.

BITE 25

SUGAR

(*Saccharum officinarum*)

Columbus brought oxen and other domestic animals to the New World on his second voyage in 1493. He also brought sugarcane, which would have such an enormous impact on the Americas that the ramifications are still being felt today. The Spanish and the Portuguese recognized that sugarcane in the New World could reap enormous profits. It required a tropical climate—not a problem in the West Indies and Brazil. It also required backbreaking work in the fields and in the mills. Europeans turned to importing enslaved African captives to do that. The Caribbean colonies of the French, British, Spanish, and Dutch eventually took on the romantic label of the "sugar islands," but the brutal labor conditions and sky-high mortality rates of the slaves belied the sweet-sounding name.

People have cultivated sugarcane for more than ten thousand years, starting in New Guinea. The crop gradually moved westward through southern China. The manufacture of sugar crystals from cane juice first appeared in India in 350 AD, and spread to Persia by the fifth century. The cane and its heavenly product of sugar followed the spread of Islam, and the Muslims introduced it to Europe.

The rapid expansion of this harsh plantation economy did lower the price of sugar and increased coastal trade among the colonies. America developed a remarkable sweet tooth as a result. Sales of molasses, another sugarcane product, also boomed. Colonists sweetened baked goods, porridge, and beverages with molasses. Unlike sugar, which has a purely sweet taste, molasses added flavor and kept baked goods moist. Plus, it was much cheaper than the cone-shaped loaves of white sugar, which, like white flour, was a luxury product. Today we know molasses added valuable vitamins and other nutrients to people's diets. (It would be key to the production of rum—see Bite 28.)

Parliament first taxed the American colonists' profound taste for sugar and molasses in 1733, in an attempt to help cover the costs of governing its far-flung territories. We Americans more or less ignored this tax. It was too high to be realistic, smuggling into the inlets and coves of the American coast was far too easy, and there was little enforcement. Later in the 1700s, Parliament got serious about taxing sugar and molasses. Britain had just won an expensive war against the French—known as the French and Indian War here and as the Seven Years' War in Europe—which ended in 1763. And it needed cash to fill its depleted coffers.

The Crown believed that the American colonies should help pay off the war's debt, since the British had outlaid considerable sums to protect the frontier settlements. So in 1764, the British revised the tax, lowering it by 50 percent, but boosting the enforcement.

This did not sit well with the colonists. Sugar became more than a luxury sweetener. It morphed into a political issue, becoming a symbol of taxation by the British government and a rallying point for American colonists chafing under the restraints imposed by the British Empire.

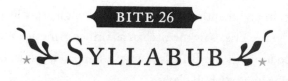

BITE 26

✿ SYLLABUB ✿

While sugar and its tax grew as an irritating political issue between Britain and the colonies in the 1760s, it was still being enjoyed as part of a fashionable drink consumed by Americans and British alike.

Few words are as evocative as syllabub, even if you don't know what it means. A syllabub is not a low-brow comedic poem or a weird version of the word "syllable." It is a frothy beverage made of whipped cream, sugar, flavorings, and wine or brandy. Modern eggnog provides a contemporary equivalent, but eggnog is consumed at holiday parties—and often sluggishly at that. The British enjoyed syllabub all year long.

First recorded in the 1500s during the Tudor dynasty, it remained popular for more than three hundred years. In the eighteenth

A syllabub glass.

century, it was chic and sophisticated, served at the best entertainments. Syllabub reached the acme of fashionable comestibles in Great Britain and its American colonies. It had its own glassware, its own utensils, and its own unique apparatus to produce the perfect froth. In many ways, syllabub illustrates what historians call the "consumer revolution" of the eighteenth century, when middle-class households began to acquire manufactured goods on an unprecedented scale.[5]

Of course, people had consumed syllabub from earthenware mugs for generations. But now something new was available, aspirational, and almost affordable. The specially designed glassware showed off the billowy layers of wine, rich milk, and whipped cream. Merchants advertised their collections of the glasses and accoutrements for sale at their establishments.

Colonists in eighteenth-century America bought this luxury glassware at a surprising rate. They saw themselves as British Americans and felt compelled to keep up with the British Joneses in apparel, foods, house decorations, and items like syllabub glasses.

Syllabub remained popular throughout the Revolutionary era, even without the fancy glasses, until it eventually fell out of fashion. But judging by the number of recipes available online today, it seems to be making something of a comeback. The following recipe is similar to one belonging to the family of founding father and fifth U.S. president James Monroe.

Many colonists fell into debt to British agents as they rushed to purchase newly available consumer items. Ultimately, their debt contributed to the rising anti-British sentiment and led to patriotic covenants to "Buy American"—known to historians as nonimportation agreements.

During India's fight for independence from Great Britain after World War II, the pacifist leader Gandhi borrowed a page from our book and urged similar nonimportation agreements in India to limit dependency on British-manufactured goods. Today, some people use the Buy American slogan as a rallying cry to protect our national economic interests in the global marketplace.

COLONIAL SYLLABUB

This syllabub recipe would be the same in England and in the British colonies. You can use a hand whisk to be truly authentic when you whip the cream, but I stick to an electric mixer. Either way, the syllabub benefits from a sprig of rosemary in each glass.

INGREDIENTS

- 1½ cups heavy cream
- ½ cup confectioners' sugar
- 2 tablespoons finely grated lemon peel
- Juice of 2 lemons
- ¾ cup white wine or dry sherry, or a mixture of the two equaling ¾ cup
- 6 sprigs rosemary

DIRECTIONS

Pour cream into the bowl of an electric mixer and beat on medium until the cream holds soft peaks. Add the sugar, lemon peel, and juice of the lemons, with all seeds removed. Beat these ingredients briefly, and pour in the wine or sherry. Continue beating on medium for about 3 minutes. Pour the mixture into parfait glasses or wineglasses. The mixture will separate. Refrigerate overnight and serve with a sprig of rosemary in each glass.

Makes about 6 servings.

BITE 27

❧ PIE ❧

Sugar continued to infiltrate everything, from drinks and snack foods like waffles and doughnuts to the inimitable pie. Americans did not invent pie, but they ate a lot of it, serving it as a savory dish and a sweet with breakfast, midday dinner, and supper. Following medieval practice, the earliest settlers used dense pie crust to create a "coffin" (as previously mentioned in the eel pie recipe), a container in which to cook seasoned meat or fish dishes, sort of like a casserole dish we'd use today. The pastry coffin, made of whole wheat flour and lard, might have been tough and unappetizing, but it was served as a disposable, sealed container rather than a tasty, flaky crust.

Pastry skills joined other talents, such as sewing and housekeeping, as important domestic arts. Pie baking and a woman's marriageability knotted together, as illustrated by a humorous folk song from a later date, "Can she make a cherry pie, Billy Boy, Billy Boy?"

As the 1600s progressed, the idea of elegant pastry blossomed in France and influenced the style of pie crusts in Britain and her colonies. Technology gave bakers a push too. Wall ovens, built into the sides of brick hearths, made baking easier and more predictable than cooking with lidded pots over and under hot coals. Today we would find hearth-heated ovens challenging and prodigiously time-consuming, but compared to open-hearth cooking, this was a big step forward.

Sweet pies varied throughout the colonies based on local produce. Double-crusted blueberry pies starred in northern New England. German settlers, called the Pennsylvania Dutch by their Anglophone neighbors, adapted the English treacle tart and created the indigenous shoo fly pie made with molasses. The southerners often presented my favorites—peach pie, chess pie, and rich pecan pie, always crowd-pleasers.

It's worth examining the evolution of apple pie, something seen as wholly and definitively American. But despite our devotion to the humble apple pie, that's not quite the case. Here's a quick deconstructed history of this quintessentially American dessert. Apples, as noted previously, traveled across the Atlantic with the first settlers; they're not native to the Americas. The phrase "As American as apple pie" is only misleading to those who forget that, except for American Indians, we are all transplants on this continent, just like apple trees. In fact, not a single ingredient in apple pie originated in the land now defined as the United States. Butter, lard, wheat flour, apples, sugar, cinnamon, allspice, and nutmeg, even a hint of vanilla, are all imports—just like Americans (except for American Indians, of course).

An illustration of a colonial family preparing food together. The woman to the right is rolling out pie crust.

SHOO FLY PIE

I attended the Moravian school in Bethlehem, Pennsylvania, where the little cafeteria in the basement lunchroom occasionally served the marvelously named Shoo Fly Pie. At home, I add chopped walnuts to the crumble on top, which is not exactly authentic but I like the texture. Colonial cooks were less apt than we are to follow "receipts" word for word, partly because most recipes were not written down and also because a little experimentation was perfectly okay.

Modern recipes for this pie use light molasses. I use dark molasses here for a more traditional taste. For another touch of authenticity, use a lard-based recipe for the crust. Incidentally, the name "Shoo Fly Pie" was not recorded until the nineteenth century, but the Pennsylvania Dutch served a molasses-based pie in the 1700s.

INGREDIENTS

CRUST

- ▶ 1 unbaked pie shell

CRUMBS

- ▶ ½ cup brown sugar
- ▶ ½ cup unbleached all-purpose flour
- ▶ ½ cup walnuts, chopped
- ▶ 2 tablespoons cold butter
- ▶ ½ teaspoon salt

FILLING

- ▶ ¾ teaspoon baking soda
- ▶ ¾ cup boiling water
- ▶ 1 cup dark molasses
- ▶ 1 whole egg, beaten
- ▶ 1 teaspoon vanilla extract
- ▶ ½ teaspoon ground ginger
- ▶ ½ teaspoon salt
- ▶ ¼ teaspoon ground cloves
- ▶ ¼ cup chopped walnuts for topping

DIRECTIONS

Crust: Preheat oven to 425°F. Make pie crust, using your own recipe. Line a pie pan with the crust to create a pie shell. Bake crust unfilled for 10 minutes, using pie weights if you have them. Remove the pie shell from the oven.

Crumbs: Place the sugar, flour, walnuts, butter, and salt into the bowl of a food processor, using the metal blade. Process until crumbs form. Remove ⅓ cup of the crumbs, which will serve as the topping. The rest of the crumbs will be stirred into the filling.

Filling: Preheat the oven to 350°F. Place the baking soda in the middle of a mixing bowl and pour in boiling water so that the soda will combine thoroughly with the other ingredients. Add the molasses and whisk to mix thoroughly. Add the egg, vanilla, ginger, salt, and cloves. Now fold in the crumbs, except for the reserved ⅓ cup. Pour this mixture into the pie shell.

Mix the remaining crumbs with the chopped walnuts. Sprinkle the crumb and walnut mixture on top of the pie. Bake for 45 minutes.

Serves 8.

The pie may puff and start to crack a little as it bakes. This is fine; Shoo Fly Pie is not about perfect beauty. It's about homemade deliciousness.

RUM AND WHISKEY

While the ingredients for apple pie did not originate in America, two types of distilled spirits—rum and bourbon whiskey—have truly New World roots.

On a per capita basis, colonists ate more pie, and they also drank more alcoholic beverages than we do today. Their near-constant consumption of spirits, from mild perry to strong brandy, startles modern-day history explorers. John Adams, as a young lawyer in Massachusetts and later the second president of the United States, started every day with a mug of hard cider, and his contemporaries considered him a very sober man. Other people drank to excess; Benjamin Franklin collected over two hundred words to describe the inebriated state, including "addled, boozy, and buzzey."[6]

> On Election Day, candidates bought rounds of whiskey shots for their constituents to encourage them to vote in their favor. Reminder: Politics was a white-men-only game at this point.

Colonial Americans found great delight in liquor-laced punch. Rum, based on sugarcane juice, provided the foundation for various punches consumed by both genders, all social classes, and a surprising range of ages. (There was no legal drinking age of eighteen or twenty-one back in those days.) Rum is still popularly associated with pirates, sailors, and the navy, quite legitimately, because it was served on board large ships and small schooners up and down the Eastern Seaboard. In fact, it provided an important foundation for America's maritime trade.

A steady stream of spirits flowed from the distilleries in New England all the way to the West Indies, home of the sugarcane. From there, molasses traveled north to be distilled into rum, and the process started all over again. Rum also crossed the Atlantic, as one leg of the triangle trade, which included shipping enslaved Africans as part of the return voyage. (Frankly, that triangle was pretty raggedy, because sea captains covered a variety of routes, but it is a helpful way to envision the Atlantic shipping community.) By the early 1700s, rum was the most popular distilled beverage in the colonies.

Whiskey was a newcomer compared to rum. Consumption of whiskey grew slowly in the eighteenth century. But at the midcentury point, a quarter of a million Scots-Irish immigrants arrived and contributed their Old World experience with whiskey making. Local grains—corn, rye, and wheat—provided the mash for local distilleries.

A barrel of whiskey brought a higher profit margin than a barrel of flour. It did not go rancid or get infested with mealy worms, either. Enterprising farmers adopted whiskey production as a way to improve their income.

Faced with a mountain of debt after the Revolution, the federal government levied an excise tax on whiskey producers. In 1794, during the second term of Washington's presidency, the Whiskey Rebellion led by Pennsylvania farmers would challenge the federal government's right to tax individual citizens.

Maybe President Washington grew intrigued with the whiskey business at the time of his rebellion. His Scottish manager, James Anderson, encouraged his employer to build the largest distillery in America, using a mix of rye, corn, and 5 percent malted barley. In 1799, at its peak, the Mount Vernon distillery produced eleven thousand gallons and contributed significantly to the family coffers. To paraphrase a quote from Light Horse Harry Lee's famous eulogy of our first president, George Washington: first in war, first in peace, and first in whiskey.[7]

Both rum and whiskey remained very popular spirits throughout the following century. As we will see in the next chapter, in the 1800s, the truly American bourbon grew more popular than the traditional blended rye that our first president had produced with such remarkable success.

❧ TEA AND COFFEE ↴

It's true: Americans consumed a lot of liquor in the past and continue to enjoy alcoholic beverages today. However, we also are a nation of gung-ho coffee drinkers. Today, coffee accounts for 83 percent of all hot beverages consumed, with black tea, green tea, and hot chocolate sharing the measly remainder of 17 percent. It was not always thus.

The English mania for drinking tea spread like wildfire across the Atlantic around the 1720s. British merchants shipped the black shriveled leaves all the way from China. Market demand meant that they could charge a pretty penny. And along with tea arrived the fashionable accoutrements: the teakettle for boiling water, the teapot for brewing, the tea strainer, the tea linens, the sugar bowl, the tongs, the little teaspoons, and of course, the dainty teacups—a vast array of consumer items to display and enjoy. Like syllabub glassware, tea things illustrate the rise of the consumer revolution in colonial America. In some colonies, 50 percent of the households owned a tea set by the 1750s. Americans loved their tea as much as the British did.

That's one reason why people did not respond well when Parliament granted the British East India Company a monopoly on the American tea trade. The tea wasn't more expensive, but enforcement provided for a different court system and a harsher attitude toward smugglers. When a group of colonial activists threw a shipload of tea into Boston Harbor in 1773, the heavy-handed response of the British government led directly to the Battle of Lexington and Concord and ultimately to the American Revolution.

Ironically, the leaders of the Boston Tea Party in 1773 planned their political demonstration at a coffee house called the Green Dragon. Coffee was not a new addition to colonial beverages—Dorothy Jones of Boston became the first licensed coffee trader in America in 1670. Coffee really took off in New York City, even replacing beer as the preferred breakfast drink in the early 1700s.

The rejection of tea and other imports symbolized the rejection of British dominion. Americans looked to raspberry leaf and other locally grown herbal teas to replace the Bohea leaf sold by the British East India Company.

Once imported tea became politicized as a drink fit only for loyalists to the Crown, it dropped out of fashion. Tea drinkers were criticized by their neighbors, and a new age of coffee drinking dawned. Grown in the New World, coffee did not represent British economic interests. It was hot and highly caffeinated, and it retained much of its popularity even after the Revolution, when tea drinking no longer made one a pariah.

Many Americans returned to tea drinking when the war ended in 1783, just as they happily returned to purchasing British manufactured goods. Middle-class families embraced the ritual of afternoon tea in the nineteenth century, demonstrating the domestic virtues associated with genteel tea service. Coffee houses thrived in East Coast ports, where merchants and other (mostly) men met to discuss business, and had its place as a pick-me-up. It should be noted that the former French and Spanish colonies also maintained a coffee-drinking tradition. But tea upheld its position in American culinary culture, just as Americans continued to look to Britain for other cultural trends such as fashion, literature, landscaping, and art.

YAUPON TEA

Catawba Indians in what is now the southeastern United States made their own tea from yaupon holly, *Ilex vomitoria*. This holly tree is one of the few indigenous plants in the New World that contains caffeine. The early Spanish settlers enjoyed the yaupon tea and exported it to Europe for a short time. The Latin name *vomitoria* gives you warning that too strong a cup of this brew has emetic qualities, leading to a badly upset stomach.

So when did we become a nation of devoted coffee drinkers? Coffee would completely eclipse hot tea in 1865, when Union soldiers trooped home from the Civil War. The U.S. government had issued coffee as part of their standard rations, and returning veterans kept right on drinking it. By the twentieth century, people drank coffee everywhere in the United States. And today, coffee and its myriad variations are found from Seattle street corners to rest stops along Route 95. America runs on coffee.

BITE 30

GREEN PEAS

Revolutionaries sought more than political independence from Great Britain. They looked for improved ways to publicize the benefits of the New World. While fashion followers would continue to look across the Atlantic for imports and inspiration, American farmers and planters hoped to inspire the best in agricultural techniques, leading by example.[8]

The gorgeous vegetable garden at Thomas Jefferson's Monticello.

George Washington carefully chose native trees to plant around Mount Vernon and spent considerable energy improving animal breeds—from foxhounds to hogs—even while president. Thomas Jefferson, a lifelong connoisseur of food and wine, failed at establishing viticulture and olive orchards in Virginia. But his stunningly beautiful and productive vegetable garden at Monticello encouraged the adoption of "new" produce for the American pantry.

Jefferson's lengthy stay in France and his travels through Europe opened his eyes and mouth to new foods. He helped popularize the ironically indigenous tomatoes in America. Once wicked "love apples," they were now suitable for President Jefferson's guests at the newly built White House during his administration from 1800 to 1808.

European peasants found that peas dried well, providing valuable vitamins and protein over the winter. These traditional field peas were starchy and pale, more like a chickpea, and formed the basis of the "pease porridge hot, pease porridge cold" mentioned in the nursery rhyme.

Vegetables were no longer limited to the old kitchen garden standbys of carrots, cabbages, and turnips. Green beans, fresh sweet corn on the cob, ruby red beets, and glistening eggplants, not new but newly popular, expanded the list of local vegetables. And they were no longer boiled to death before being served, which helped improve their flavor and appeal. (Trust me, if you ever had cabbage cooked to death, you would turn up your nose at it too.)

Virginia cookbook author Mary Randolph advised her readers not to overcook beans, adding, "As soon as they are tender, take them up, and throw them into a colander to drain."[9] And sometimes, vegetable gardeners even harvested at least part of their crop when young and tender to enjoy immediately, as was the case with green peas.

I have all sorts of reasons to like green peas. They have a very long history, for one thing. Archaeologists have found evidence that people were eating peas more than eight thousand years ago. In ancient Athens, street vendors sold takeout hot pea soup. (Yes, takeout food is not the modern convenience you thought it was!) The appreciative Romans named this legume *pisum* (pronounced pee-sum) from which we get the word "peas." And peas served as an important source of nourishment for medieval Europeans.

The tasty little green peas people enjoy today hit the big time in Renaissance France, where King Henry II's Italian wife, Catherine de Medici, introduced them in the 1500s. There, they became known as *petit pois*. By 1695, a lady friend of Louis XIV, the famed Sun King, described the excitement at court over the first fresh peas harvested that season as "a fashion, a craze!"[10]

It's hard to imagine that baby peas could inspire such passion. But after his presidency, Thomas Jefferson indulged in some healthy competition with his Virginia neighbors over who could harvest the first green peas of the season. He was thrilled when his garden at Monticello produced the first crop in the neighborhood. That's a lot of excitement over a small green legume for one of our most cerebral presidents. Because peas ripen early in the summer, they were among the first bright vegetables on the table each year throughout the millennia before refrigeration.

Other reasons to like green peas: they can be planted early—around St. Patrick's Day in the mid-Atlantic region (Zone 7)—and harvested early as well, as mentioned above. Also, they freeze really well and retain their taste, texture, and nutrients when frozen commercially. So unless you are lucky enough to eat peas straight from a vegetable garden, baby peas from your grocer's freezer are as close to fresh as you'll get, and they are delicious.

Today I look for the frozen tiny baby peas at the supermarket that are just a fraction smaller than the so-called garden peas. They are sweeter and less starchy and hardly need to be cooked. I leave the package in the fridge overnight so they defrost slowly, and then I just heat them through for about one minute in boiling water before draining. It's traditional to add a bit of chopped mint to fresh peas, but a pinch of rosemary is tasty too.

chapter 4

ROAST TURTLES
AND HANGTOWN FRY

The Rise of a New Nation

ixty years is a vast period in United States history, and the decades between 1800 and 1860 prove no exception to this rule. Starting as an energetic, adolescent nation, America would stretch across a continent, form a unique cultural identity as a nation of immigrants, and wrestle with the paradox of slavery in a country dedicated to freedom.

Jacksonian democracy and economic expansion brought new opportunities for white men of all ranks. Women managed households with large families and unending workloads. Free people of color and enslaved African Americans, joined by a slowly fomenting antislavery movement particularly in the northern states, pursued the hope of liberty.

Along the way, a rich variety of foods and culinary traditions developed, highlighting the truth in our national motto. *E Pluribus Unum*—One from Many—describes the individuality of the regional foods explored in this chapter as they weave into an American gastronomy the same way many people from all over the world form the colorful fabric of our diverse culture.

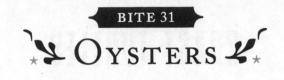

BITE 31
OYSTERS

Oysters were America's first fast food, popular long before the first colonists came to stay.

Shell heaps, called "middens," along the Atlantic and Pacific coasts stand in mute testimony to the American Indian love of oysters. Archaeologists have found middens dating from 2000 BC along the Louisiana coastline, indicating that American Indians were enjoying oyster feasts before the first Olympics in ancient Greece. Early European explorers wrote of the abundance of this mollusk, which was easy to harvest, nutritious, and a healthy indulgence particularly in the cold weather.

In colonial America, street vendors sold oysters, hawking their wares in seaports from New Hampshire to Georgia. Raw, roasted, fired, stewed, chowdered, or baked in a pie, the shellfish appeared on linen-draped tables at the most elegant establishments and plank-topped sawhorses in dirt-floored taverns. Different locales boasted of their special breeds, like Blue Points from Long Island or Olympias from the San Francisco region.

When President Thomas Jefferson completed the Louisiana Purchase in 1803, the United States acquired not only 530 million acres but some of the richest oyster beds in the world, in the Louisiana estuaries. Even today, 30 percent of the oysters consumed in America are harvested in Louisiana.

Men and women harvested the mollusks right off the coast of Manhattan and sold them at fish markets or on the street, until the government closed the oyster beds due to pollution in 1926. Oysters were the fast food of the nineteenth century, cheap, easy to eat, and ubiquitous for the eight cold months of the year. (When the weather and water warmed up—in May, June, July, and August, the months without *r* in their names—there was a greater chance of bacteria affecting the quality and thus oysters weren't consumed as frequently.) And unlike most fast food today, oysters are nutritionally balanced.

Indeed, the appearance of oyster-specific utensils and china highlight the popularity of this mollusk in the 1800s. Even today, Victorian-era oyster plates are not difficult to find online or in antique stores.

For the families that own these things, great-grandmothers' oyster plates can be donated or resold without a tear being shed, since generally they haven't been used since the Eisenhower election. But oysters themselves still are just as delicious today and can be enjoyed in a number of delightful ways.

SAUCES FOR RAW OYSTERS

This is a hotly debated topic and very personal for many people. Purists insist that high-quality raw oysters benefit from a light sprinkling of lemon juice, and that's it. Others swear by a traditional French sauce, mignonette, which literally translates as "little cutie." It is made of shallots and sherry vinegar and is quite tasty. There are lots of directions online for how to make a mignonette sauce.

Although bottled red cocktail sauce is the most obvious, gourmets shudder at the thought. The sweetness can be cloying and overpoweringly wrong with raw oysters. Key lime juice, with minced fresh cilantro and a touch of cayenne, is intriguing and a much better choice. But I use chili sauce, adding horseradish and a bit of fresh lemon juice. Feel free to experiment with these delicious morsels from the sea and see what you come up with.

BITE 32

❧ Roast Turtles ☙

Oysters weren't the only bounty to be found in North America's rivers, lakes, and seas. When European settlers first arrived, turtles thronged our freshwater and coastal shores. Once again, a delicacy in Europe became an inexpensive delight in the New World. Plentiful in freshwater, estuaries, and salt water, these reptiles were easy to catch or trap.

Turtles reached their peak of fashionability in old New York, when young dandies and chaperoned young ladies would ride in horse-drawn carriages from lower Manhattan north to the farmland along the East River. There, they enjoyed turtle roasts, held at country inns situated near the water on what is now the Upper East Side.

> *American Indians devoured those big diamond-backed terrapin turtles from the Chesapeake. In fact, "terrapin" is an Algonquian word.*

Although turtle roasts faded in popularity as social events, perhaps because the turtle population in the East River declined, turtle soup maintained its popularity. There are many old recipes, several of which acknowledge the difficulty of prying open the carapace, or shell. It's not common in our national cuisine today, but turtle soup can still be found on menus in the Chesapeake region and Louisiana. You can buy the frozen meat online if you want to try your hand at cooking it yourself.[1]

BITE 33
Ice Cream

In the early 1800s, roast turtles found a new degree of fashionability, and other food trends were on the rise as well. Dolley Madison, First Lady par excellence, served ice cream at the White House, and suddenly ice cream was everywhere. And unlike roast turtles, ice cream remains one of America's favorite foods.

> *George Washington acquired a taste for ice cream while living in New York during his presidency and spent $200—about $4,600 today—on the frozen treat while entertaining guests in the summer of 1790. That's a lot of ice cream! He also bought a cream machine for ice for his home at Mount Vernon.*

Culinary historians debate when the first "iced cream" appeared, but the late seventeenth century seems like a safe bet. We know that the ancient Romans enjoyed fruit syrup with crushed ice or snow—the infamous Emperor Nero sent teams of slaves running to the Alps for fresh snow for his drinks. "Ices" were popular in the Renaissance courts of France and Italy as well. But true ice cream is a dairy product, churned while it freezes, not a slushy fruit drink. And the 1600s seem to have witnessed its arrival on the culinary scene.

By the mid-1700s, ice cream appeared at elegant gatherings in colonial America. In Maryland in 1744, provincial governor Thomas Bladen served it to his guests: "…a Dessert no less Curious; Among the Rarities of which it was Compos'd, was some fine Ice Cream which, with the Strawberries and Milk, eat most Deliciously."[2]

Today we often associate ice cream with the gregarious and hospitable Dolley Madison. Wife of the brilliant but reserved fourth president, James Madison, Dolley served ice cream at his second inaugural ball during the War of 1812. It became "her" dessert for White House dinners.

Mrs. Madison hosted glittering dinners to advance her husband's political agendas, inviting people with different political views to share in abundant wines, meats, and of course, the ice cream. At the time, ice cream was labor intensive and required a plentiful supply of chipped ice, which was an extravagance in the warm weather.

An ice cream maker.

In 1832, an African American cook who had worked at the White House, Augustus Jackson, invented a prototype for the churn used to manufacture ice cream. Although he failed to patent his device, he ran a popular ice cream parlor outside Washington, DC. Eleven years later, another unsung inventor, Nancy Johnson, patented the hand-cranked ice cream freezer which remains the basis for subsequent ice cream makers, or "artificial freezers," as she called the device in her patent in 1843.

Ice cream comes in many formats and flavors today, from the honestly rich dairy ice cream to nonfat frozen yogurt and even soy- and almond-based ice creams, and from plain vanilla to chocolate salted-caramel with bacon. We like to think we invented this concept of exotic flavors. But no one comes close to Dolley Madison, who liked to tuck into a delicate dish of oyster ice cream.

BITE 34

❧ BRUNSWICK STEW ❧

Historical records reveal Dolley Madison's love of ice cream, but she is also famous for her heroic actions when the British burned the White House during the War of 1812. Mrs. Madison saved irreplaceable works of art (a life portrait of George Washington, for example) and documents (the

Constitution!) through her quick thinking. In many ways, the War of 1812 was the finale to the American Revolution, permanently removing the British military force from our Northwest Territories, comprised of current day Illinois, Indiana, Michigan, Ohio, and Wisconsin.

The end of the war in late 1813 also opened up new territory for people in search of their own land. Now that the British had left, more families followed trailblazers like David Crockett westward looking for their own homesteads. The deep forests that stretched to the Mississippi echoed with the sound of ax blades hitting tree trunks as people cleared the land for farming.

The pattern of hardship and rustic living on the frontier of the Eastern Seaboard repeated as the frontier itself moved westward, from Pennsylvania to Ohio and on to Illinois, or from Georgia to Alabama to Mississippi. Families lived in small cabins until, little by little, they might add an extension, nail wooden planks over the dirt floors, and even build a chimney with the help of a skilled neighbor.

The tough childhood of Abraham Lincoln, born in 1809, is one well-documented example of growing up in pioneer days, a label overflowing with romantic images of coonskin caps and sturdy fellowship. But it was hard living, particularly for the young wives who moved westward with young children and, often as not, another on the way. Even when the cabin was built and some land was cleared, getting meals on the roughhewn plank table was burdensome.

> *Several locales claim to be the source of Brunswick stew, including Brunswick, Georgia, and Brunswick County, North Carolina. The late southern food writer John Egerton pointed out that American Indians made stews with wild game, beans, and corn long before any place in North America was named "Brunswick," so they should get the credit as originators.[3]*

One of the most efficient ways to feed a big family was to cook one-pot meals over hot coals (a tradition that made a comeback with the invention of slow cookers in the late twentieth century). That was the way American Indians prepared many of their meals, and it worked for frontier families as well. Like a modern stew, a little variation in heat or ingredients turned out just fine as long as the stew didn't burn.

Brunswick stew is a classic dish from days of the early republic. Early recipes call for squirrel meat, but poultry works well too. Just like those hardy women in their cabins, you can substitute what you have on hand—sausage, pork, rabbit. I do think some corn is required, as well as lima beans. Today the various printed recipes always include tomatoes—I would guess those showed up later in the nineteenth century. And my version starts with some smoky bacon or ham, because don't you think our foremothers would have stuck some bacon in the pot if they had it on hand?

BRUNSWICK STEW

INGREDIENTS

- ► 6 pieces thick-cut smoked bacon, diced
- ► 2 young chickens, about 3 pounds each, cut into pieces
- ► 2 squirrels, dressed, cut into pieces
- ► 8 cups chicken broth
- ► 3 medium potatoes, peeled and chopped
- ► ½ cup chopped onions
- ► 5 cloves garlic, chopped
- ► 1½ cups corn, fresh or frozen
- ► 1½ cups lima beans, fresh or frozen
- ► 1 cup peeled and chopped tomatoes
- ► 1 tablespoon crushed or powdered sage
- ► 1 teaspoon dried thyme
- ► Salt, cayenne, and freshly ground black pepper to taste

DIRECTIONS

Sprinkle the bacon into a large frying pan at room temperature. Place the pan on medium heat, turning the bacon occasionally until crispy. Remove the bacon from pan, drain on a paper towel, and reserve. Brown chicken and squirrel pieces in batches in bacon fat. Place the browned meat pieces in a large stew pot or Dutch oven. Cover with chicken broth, and simmer gently, covered, for 1 hour. Remove meat pieces from the pot with a slotted spoon or tongs, set aside, and let cool.

Meanwhile, cook potatoes, onions, and garlic in remaining bacon fat in the large frying pan until onion is transparent, and reserve until the cooked chicken has been removed from the pot. Add the potatoes, onions, and garlic to the pot. Simmer, uncovered, for 30 minutes. Add the meat pieces and reserved bacon to the pot, along with the corn, lima beans, tomatoes, sage, and thyme. Cook for an additional 10 minutes, keeping the heat low. The potatoes should be quite tender. Add salt, cayenne, and black pepper.

Serves 12 to 14 as a main dish.

If you are using squirrel, do not include the brains. One rabbit, cut into pieces, can substitute for the squirrel. It's traditional to leave the bones in the meat, but suit yourself. For the tomatoes, I use Muir Glen Organic fire-roasted crushed tomatoes. This stew is even better served the next day.

BITE 35

⚘ CAKE AND THE ERA ⚘ OF ANDREW JACKSON

Andrew Jackson, gambler, dueler, and frontier general, lived in Tennessee, one of the young states in the West where a Brunswick stew might simmer over coals in a rustic cabin. Jackson's treatment of American Indians will tarnish his legacy forever, although he remains a hero to many. During his time in Congress and as president, more men were allowed to vote than ever before. Election days were big celebrations. Often, a lot of liquor was involved. And a big cake would be there too. This is really the story of celebrations, for cake featured prominently at happy events and large gatherings in early America.

Today, a good-quality boxed cake mix with added fresh ingredients reliably produces an excellent dessert in no time even in the hands of a novice. But for most of history, cake baking was a time-consuming art form and, rendered deliciously, represented diligence and exceptional culinary skill in nineteenth-century America. An Election Day cake during the age of President Andrew Jackson (1829–1837) would have been particularly impressive and large enough to feed a crowd.

There are lots of reasons for cake's revered status throughout the Age of Jackson. Leavening—the ingredient that makes baked goods, like cakes, rise—presented one challenge. Today we simply buy baking powder at the store. A precise blend of baking soda and cream of tartar, it's easy to measure, inexpensive, and reliable. But there was no commercially available baking powder in the early 1800s.

To make a cake rise, a cook used either eggs for a light sponge cake, yeast, or pearl ash. None of these processes was easy or predictable. The eggbeater had not yet been invented, and yeast varied in its leavening power and taste. And producing pearl ash, a homemade baking soda, was a dirty job with a discouragingly unreliable result. (Commercial baking powder, invented in Great Britain in 1845, was not sold in shops until the 1850s, when it changed cake baking forever.)[4]

Because a cook had to wait for the yeast to work its magic and raise the batter, cake making was time-consuming. Most recipes call for large quantities of ingredients. No one seems to have ever made one small yeast cake, just the way no one baked one small loaf of bread. Once you undertook the culinary challenge and time commitment of cake baking, you might as well go all the way size-wise. Martha Washington's cake recipe called for five pounds of flour and forty eggs. (Of course, her talented enslaved cooks—Doll, Hercules, and later, Lucy and Nathan Lee—actually baked this fine dessert.)[5]

Another challenge to the American cook was the oven. It took an even source of moderate heat to bake a cake, and there was no absolute guarantee of that in many home kitchens. Those without wall ovens built into the hearth could buy tin ovens, which reflected the fire's heat for roasting as well as baking. Still, keeping a fire sustained at a certain heat required carefully monitoring. A child, servant girl, or young slave might be harshly reprimanded (or whipped) for allowing a fire to go out or reach too high a heat, ruining the cake.

TIN-KITCHEN REFLECTING OVENS AND WOODSTOVES

Reflector ovens became popular in the late 1700s, and settlers brought them in their wagons as they moved westward. Made of shiny tin, which reflected the fire's heat, they are still manufactured today and sold to hardy souls who are living off the grid or simply experimenting with open-hearth cookery.

Ben Franklin invented his famous stove in 1742, but that was designed to heat a room efficiently, not cook a dinner. By 1800, large cookstoves were used at military forts and the great houses of Europe to feed literally armies of people. Philo Stewart patented the Oberlin stove, a wood-burning cast-iron stove sized for domestic use, in 1834, and it was very popular, with ninety thousand units sold over the next thirty years.[6] Still, many more households used tin ovens until well after the Civil War, because they were much cheaper and about the size of a bread box. In some communities, women would bring items to be baked to the local baker, where, for a fee, the raw mixed ingredients would be cooked in professional ovens. In the Southwest, outdoor community ovens might be shared by neighbors.

Ladies proudly served up big yeast cakes with hard icing at weddings, funerals, holidays, church fairs, and election time, when people gathered together for speeches and voting. Of course, voting was limited to white men at this time. Andrew Jackson, rightly famous for extending the suffrage to the common man, did not advocate for gender or racial equality, and would have been out of step with his era if he had. Still, whole families traveled into towns or villages so that their men could cast their votes. With out-of-town relatives and friends gathering, Election Day was a social occasion, and a festive one at that, a day that cried out for a nice, big cake to feed a crowd.

ELECTION DAY CAKE

This recipe for a traditional Election Day cake comes from the second edition of Amelia Simmons's American Cookery, *the first cookbook written by an American, a historical event in itself.*

ELECTION CAKE

Thirty quarts of flour, 10 pound butter, 14 pound sugar, 12 pound raisins, 3 doz eggs, one pint wine, one quart brandy, 4 ounces cinnamon, 4 ounces fine colander seed, 3 ounces ground allspice; wet flour with milk to the consistence of bread overnight, adding one quart yeast; the next morning work the butter and sugar together for half an hour, which will render the cake much lighter and whiter; when it has rise light, work in every other ingredient except the plumbs, which work in when going into the oven.

EARLY COOKBOOKS

Cookbooks are not a modern invention. Our first English settlers followed recipes similar to those found in the English cooking manual by Gervase Markham, who in 1615 published the fabulously titled *The English Hus-wife, Containing the Inward and Outward Virtues Which Ought to Be in a Complete Woman*. *The Art of Cookery Made Plain and Easy* (1747) by English author Hannah Glasse served a large audience on both sides of the Atlantic, and stayed in print for over fifty years. Amelia Simmons's *American Cookery*, originally printed in Hartford, Connecticut, in 1796, went through thirteen editions by 1831. Simmons authored the first cookbook to combine several American ingredients with English culinary techniques. For instance, *American Cookery* contains the first known recipe for serving turkey with cranberries and the first printed use of cornmeal in a johnnycake "receipt" (recipe). For those of you who enjoy literary comparisons, Simmons was the James Fenimore Cooper of cookbooks.

Mrs. Mary Randolph published her popular *The Virginia Housewife* in 1824, with carefully organized recipes, including a "fricassee of cods' sounds [bladders] and tongues" and advice on making rose brandy. Many of her "receipts" foreshadow fashionable dishes today, such as gumbo and polenta, as well as the homey but always popular baked macaroni and cheese. This may be the first cookbook in the United States that mentioned pasta.[7]

BITE 36

❧ SPANISH CALIFORNIA ❧ RABBIT STEW

Beginning in the 1830s, during the Age of Jackson, wagon trains began heading west, starting a vast migration of individuals and families from the

East Coast to the Pacific Northwest. And wherever women traveled, they brought their family "receipts"—for big election cakes, for chicken pie, and for preserves—with them. By and large, the families settled from Missouri west to Oregon, because all of California belonged to Mexico.

With encouragement from President James Polk, Congress declared war on Mexico in 1846. Two years later, the Mexican War ended. The United States expanded its boundaries "from sea to shining sea" through the invasion, conquest, and purchase of Mexican territory north of the Rio Grande as well as California. A rich food culture combining New World food with Spanish traditions soon became part of American regional cuisine. (Although few easterners would consider tamales a traditional American food, they are perhaps more American than apple pie, which as we've learned is made entirely of imported ingredients. The ingredients for traditional tamales come from the Americas.)

A BRIEF HISTORY OF EARLY CALIFORNIA

Franciscan missionaries had traveled north along the California coast in the 1700s, converting American Indians to Christianity and introducing sheep and cattle ranches as well as productive vineyards and orchards. Led by Spanish priest Fra Junipero Serra, they built a chain of Catholic missions along the coast which provided centralized hubs for worship, farming and ranching, trade, security, and healing. Native peoples, many of whose families were decimated by European diseases, were forced to assimilate and resettle around the missions. Misión San Francisco de Asís, founded in 1776 on a hilltop overlooking what is now called San Francisco Bay, presents a fascinating example of one of the twenty-one historic missions in California that still stand. Now known as Mission Dolores, it lends its name to the colorful and diverse Mission District in San Francisco.

The Franciscans were followed by prosperous families of largely Spanish descent, who originally claimed extensive lands in California granted by the Spanish crown and later by the Mexican government. When the United

States acquired the territory, these land grants were often ignored or challenged in court by unscrupulous settlers eager to wrest ownership from the old families. Still, some of these land-grant families were able to hold on to at least some of their acreage and their cultural traditions.

A wonderful cookbook by Encarnación Pinedo opens a window into the rich culinary heritage of Californios. Although bearing the title *El cocinero español* ("The Spanish Cook"), the recipes present an authentic mix of New World Mexican and Old World influences even when labeled Catalan, Basque, or French—testimony to the author's pride in her Spanish lineage as the granddaughter of a wealthy Creole landholder and ambivalence about the status of Mexicans in California under Anglo rule. She included two recipes for rabbit stews, tapping the plentiful supply of hares from Alta California. Pinedo finished one of the dishes with a green chili sauce, combining the tastes of the New World with the Old.

ENCARNACIÓN PINEDO'S RABBIT IN CHILE SAUCE

(*Liebra enchilada*)

Pinedo published her cookbook, El cocinero español *(The Spanish Cook), in 1898, fifty years after California's gold rush began. Including one of her recipes in this section seems appropriate because the recipe reflects a bygone era, and did so even at the time of publication. The recipe also represents her family's strong Hispanic cultural identity and her personal memory and sophisticated knowledge of food and cookery. Here are her own words describing how to cook this dish:*[8]

Cut the rabbit in small pieces. Fry in very hot fresh lard with small pieces of pork fat. Fry over a quick fire, and when it begins to brown, add some chopped onion, garlic, and salt. Let it cook. Then add tomatoes, olives, chopped mushrooms, one or two spoonfuls of flour, and powdered oregano. Cover it with a chile sauce, leaving the casserole covered, and cook it over a moderate flame.

MODERN RABBIT IN CHILE SAUCE

Note on equipment: You will need a large, deep frying pan, or a paella pan, with a tightly fitting lid.

INGREDIENTS

- ► 3 tablespoons cooking oil
- ► 1 tablespoon bacon fat
- ► 1 rabbit, cut into pieces
- ► ½ pound (8 ounces) coarsely chopped mushrooms
- ► ¼ cup chopped onion
- ► ⅔ cup diced tomatoes
- ► 4 black olives, pitted and minced
- ► 4 green olives, pitted and minced
- ► 3 cloves garlic, minced
- ► 1 teaspoon salt
- ► ½ teaspoon ground cayenne pepper
- ► 1 cup chicken broth, plus ¼ cup as needed

- ½ cup good red wine
- 2 teaspoons dried oregano
- 2 teaspoons rubbed sage
- 2 teaspoons crushed thyme
- 1 cup green chile and tomatillo sauce

DIRECTIONS

Melt oil with the fat in a large, deep frying pan or a paella pan. Add the rabbit pieces and brown on both sides over medium-high heat. Reduce the heat to low, cover, and let cook for about 10 minutes. Remove the rabbit pieces and reserve, covering with aluminum foil.

Sauté the mushrooms and onions together in the same frying pan over medium heat. Add more oil if needed, stirring frequently. Stir in the tomatoes, olives, garlic, salt, and cayenne, and cover and simmer for 5 to 8 minutes more over low heat. Add the chicken broth and wine, stir, then cover and let simmer slowly together for 5 minutes so the flavors mix. Mix in the oregano, sage, and thyme. Place the rabbit pieces back into the frying pan. Cover and cook over low heat for 15 minutes, occasionally turning the meat and coating with the sauce. Add more chicken broth if the sauce is too thick.

Serve the green chile and tomatillo sauce on the side.

Serves 4 to 6 as a main dish.

Try using cremini mushrooms for extra flavor and authenticity. Roasted potatoes and zucchini accompany this very nicely. This is really tasty, with lots of flavor layers, so it's great served with a simple salad of fresh greens and slices of crusty semolina bread. (You can find recipes for the bread online or buy it already prepared.)

HANGTOWN FRY

When Mexico sold Alta California to the United States for $15 million at the end of the Mexican War in 1848, they added much-needed cash to their depleted coffers. But the Americans walked off with the real bargain, which quickly morphed into one of the biggest bonanzas in history. For in the same year as the treaty was signed, James Marshall found a chunk of gold in Sutter's Creek. The California gold rush was on. The world beat a path to this new territory, with 50 percent of the eager hopefuls arriving by sea from Asia, Europe, and the Americas.

Hangtown is now known as much more respectable Placerville. But Mary Cory, director of the El Dorado County Historical Museum, points out that even in 1848, Hangtown was the county seat and fairly law abiding. Juries served there, and judges pronounced sentences. Hangings took place there, but followed legal process. Executions were not alcohol-fueled lynchings on a lawless frontier.

The colorful gold rush era, roughly 1848 to 1865, offers a mine of riches when it comes to all kinds of history, including some entertaining tales of food and drink. Mark Twain, for example, demonstrated a prodigious capacity for eating oysters and drinking hard liquor, feats respected by many in San Francisco during his stay there. One of my favorite stories, concerning the origins of a picturesquely named omelet called the Hangtown Fry, actually boasts two versions.

In *Twain's Feast*, author Andrew Beahrs serves up both origins of the Hangtown Fry. (A word of warning to the meticulous researcher: I can't find a documented source on Hangtown Fry prior to the twentieth century, which is too bad. But those two folk origins are the stories I could find, so I'm sticking to them.)

Both stories take place in Hangtown, California, which received its name after three desperadoes were hanged there. In one account, a newly wealthy miner arrives in Hangtown and demands the most expensive meal available.

In those early days of the gold rush, hen's eggs were as rare as hen's teeth in California, a scarce luxury and prohibitively costly. Oysters, shipped inland and uphill from the coast, fetched a high price too, although not as much as the eggs. A local cook created an omelet bulging with oysters, which cost the successful miner some of his newfound gold.

The other folk origin for Hangtown Fry describes a condemned man's request for his last meal. Wanting to postpone his encounter with the hangman's noose for as long as possible, he asked for a dish whose ingredients he knew would take several days to arrive, since the hen's eggs and oysters had to travel 130 miles overland from San Francisco to Hangtown.[9]

Whichever story you prefer, a Hangtown Fry still describes an oyster-filled omelet, with bacon strips resting like oars across the top. You can find it on quite a few menus in El Dorado County, California. In San Francisco, the 165-year-old Tadich Grill serves Hangtown Fry at their "new" location (since 1967) on California Street, and that's where I first had a taste of this sumptuous gold rush dish. You can make it at home and eat it slowly, savoring every morsel, and postpone all sorts of things—even a trip to the scaffold.

HANGTOWN FRY

Tadich Grill in San Francisco serves this dish with bacon and green onions crumbled into the omelet, along with breaded, panfried oysters, which is more traditional. In my version, I warm the oysters instead of breading and frying them, which makes it a lighter dish. Even if I were a condemned desperado, I think I'd like my oysters prepared very simply—although the breading and frying does take longer!

Advice: If you have never made an omelet, I recommend going online and watching a how-to video, or asking someone to show you, because it's not as easy as it looks. Generally speaking, a bad-looking omelet tastes fine (unless you've let it burn). For a two-egg omelet, you'll need an 8- or 10-inch pan.

INGREDIENTS

- ▶ 2 strips bacon
- ▶ 3 to 4 raw oysters, shucked and drained
- ▶ 1 pinch dried thyme
- ▶ Salt and freshly ground black pepper to taste
- ▶ 2 eggs, lightly beaten
- ▶ 1 slice whole grain or sourdough bread for toast
- ▶ 1 to 2 teaspoons hot sauce

DIRECTIONS

Place the bacon in an unheated, seasoned cast-iron frying pan or Teflon pan, and fry over medium-low heat until crispy. Keep an eye on the bacon so it doesn't burn. You want the fat to be rendered and the meat browned. Remove the strips and drain on a paper towel. Pour off most of the fat, leaving just a little for flavor. Generously spray the pan with cooking spray, or add butter, to keep the surface well lubricated.

Gently put the oysters in the pan over medium-low heat. Cook about 1 minute on each side to warm them through. Do not overcook or they may get rubbery and the taste changes. They may lose a little liquid as they cook, which is fine. Remove oysters from pan and reserve.

Preheat pan to medium high. Sprinkle the thyme, salt, and pepper into the eggs. Pour egg mixture into the hot pan and distribute evenly by swirling the pan a little. Lift the cooked edges of the omelet with a spatula so that the raw egg in the center can run underneath and cook.

Place the oysters down the middle of the omelet. Loosen one side of the eggs from the pan edge with a spatula. Add butter or cooking spray to the bottom of the pan, as needed, to make this process easier. Gently fold the loosened side over the middle of the omelet. Repeat with the second side. Slide or lift the omelet off the pan and place on a warm plate.

Place the strips of bacon on the omelet, like oars on a boat, and serve. Accompany with toast and hot sauce, if you like that. I prefer Sriracha brand hot sauce.

Serves 1 but can be doubled.

If your omelet is a mess, the toast and bacon can make it look a lot better. The omelet makes a meal in itself, but from my point of view all this protein cries out for vegetables. Try a side salad of baby spinach, arugula, and some chopped apple, tossed with a lemony vinaigrette.

I get thick-sliced bacon from my butcher. This lets me pretend that it's healthy because it's hand-sliced, not prepackaged. I'm not much on shucking my own oysters, so I ask the nice guys at Jeff's Seafood, my neighborhood fish store, to do it for me. Whether you buy your oysters at the fishmongers or at your supermarket, make sure they are fresh. For this recipe, I used our local Long Island Peconic oysters, which are small and sweet, tasting like an ocean breeze at high tide.

BITE 38

❧ IRISH POTATOES ❧

Immigrants on the Overland Trail brought food supplies with them, like dried beans and potatoes. Potatoes keep well and travel well too. In fact they traveled all the way from the New World to Spain, and back again.

I hope you remember that potatoes originated in the New World, not Ireland. The Spanish brought white potatoes from Peru to Europe in the early 1500s. Certain countries, such as Poland and Ireland, adopted the potato as their own after a lengthy introductory period. By 1840, almost half of the Irish population relied on the potato as their main dietary staple. It thrived in the wet Irish climate, and neither rocky hillsides nor boggy ground limited its harvest. A quarter acre of potatoes and one milk cow could nourish a family of six.

But when the potato blight struck in 1845, famine swept across the land for the next five years. A journalist in Cork, Ireland, witnessed this: "Death is in every hovel; disease and famine, its dread precursors, have fastened the young and old, the strong and feeble, the mother and infant…"[10]

The population fell from 8.1 million in 1841 to 6.5 million a decade later, due to death and emigration to flee the famine's devastation. The Irish emigrated to Great Britain, Australia, Canada, Latin America, and the United States. Here they met intense prejudice, not only because of their "popish" Catholic religion, but also their language, accent, and poverty. In Boston, Philadelphia, New York, and other cities, mobs of Protestants violently struck at Irish neighborhoods. However, even when the blight died out and the potato harvest revived, the Irish continued to emigrate here.[11]

They were one of the first groups to arrive en masse and thus are considered one of the first "waves" of immigrants to the United States. Irish workers helped build the canals and railroads that accelerated the expansion of the economy on a national scale, particularly in the northern states. They brought their love of music, dancing, storytelling, political debate, and the tavern, all aspects of American life enhanced by their contributions.

The Irish also brought their love of the white potato. Although "native" Americans—primarily of British descent—ate potatoes much more frequently than some of their European cousins, they referred to them as "Irish potatoes" to differentiate them from sweet potatoes. By doing so, they also recognized the Celtic immigrants' devotion to this humble but nutritious tuber.

White potatoes became so commonplace that "meat and potatoes" refers to a basic American meal or really a basic, unadventurous anything, from a tour of New Orleans ("It was your basic meat-and-potatoes tour") to a medical plan. Potatoes show up at every meal—breakfast, lunch, and dinner. And nothing is more American than french fries as a snack, except maybe potato chips. What is so wonderful about these nutritious and delicious tubers is that they can be elegant and complex, like duchess potatoes, or hearty and filling, like a baked potato with butter and salt. Irish potatoes were inexpensive nourishment in the nineteenth century, and they still are today.

BITE 39
Mint Juleps

Once again I am straying from the Bite and going for the sip, and in the case of a mint julep, it is well worth it. The first record of an American iced mint cocktail is from a book by John Davis, published in 1803 in London, which describes "a dram of spirituous liquor that has mint steeped in it, taken by Virginians of a morning."[12] Long-associated with southern planters, mint juleps today are the featured cocktail at the Kentucky Derby, and in bars and living rooms across the country where horse-racing fans are indulging in spirits while cheering on their favorite to win.

THE BEAUTY OF BOURBON

People called corn whiskey "bourbon" after Bourbon County, Kentucky. Kentucky achieved statehood in 1792, about the same time that bourbon whiskey began commercial production due to the dedication of individuals such as Jacob Beam (the predecessor of Jim Beam), Jacob Spears, and the Reverend Elijah Craig. By 1820, distillers like Dr. James Crow and Jason Amburgey had standardized the sour mash process, and commercial distribution began in earnest.

Stored in large hogshead barrels, corn whiskey traveled by sea to San Francisco beginning with the California gold rush in 1849. Smaller wooden barrels filled with bourbon sat in the back of Conestoga wagons as families headed west. Particularly popular in the South, bourbon was less expensive to produce than scotch, being based on corn, the country's most plentiful grain. Modern regulations require bourbon to be 51 percent corn based, and there is still considerable variation in taste and quality among the different brands of bourbon. This variation was even more pronounced in the nineteenth century.

A mint julep is an all-American mixed drink made of bourbon, confectioners' sugar, a bit of water (or simple syrup), and plenty of mint, served over cracked or crushed ice. Whiskey replaced rum as the nation's favorite distilled liquor in the nineteenth century. The arrival of waves of Scottish Highlanders and Scots and Irish immigrants meant that trained whiskey distillers joined the ranks of immigrant Americans, and whiskey no longer had to be imported (or smuggled) into the new nation.

> Incidentally, the old English word "julep" comes from the Persian gulab, *meaning rose-water, or a refreshing drink. Not a bad name for something so cool and invigorating.*

Of course, now that the British were banished from patrolling our coast, smuggling became downright un-American and was frowned upon because it was no longer necessary. This actually presented economic opportunity for local distillers. As we noted in Bite 28, George Washington made a fat profit late in his life selling a blended rye whiskey from his impressive distillery at Mount Vernon, thanks to the expertise of his manager, James Anderson, an entrepreneurial Scotsman.

Senator Henry Clay, who rests among the pantheon of great senators in our history, introduced the mint julep to Washington, DC society, according to cocktail legend. As a Kentuckian, he was a bourbon booster, and instructed the bartender at the famed Willard Hotel to serve mint juleps one spring around 1848, when Congress was in session and fresh mint abounded in local gardens. Today, the Round Robin Bar at the Willard still serves up a refreshing mint julep, although in Senator Clay's day each cup would include several tall sprigs of peppermint for flavor, not just one decorative stem.

The classic mint julep is served in a silver julep cup, ideally with a silver drinking straw. Filled with crushed ice, the cup should have heavily frosted sides. Etiquette recommends that you hold the ice-cold cup at the base so as not to warm the drink, in case you were wondering.

CLASSIC MINT JULEP

INGREDIENTS

▶ 4 fresh mint sprigs

▶ 2 teaspoons water

▶ 1 teaspoon confectioners' sugar

▶ 6 ounces cracked or crushed ice

▶ 2½ ounces bourbon whiskey

DIRECTIONS

Strip leaves from 3 mint sprigs and put them in the bottom of your mint julep cup or an attractive 8-ounce glass. Muddle (mix together while squishing the leaves with the back of a spoon) the leaves, water, and confectioners' sugar. Fill the glass with the ice, and add bourbon. Garnish with the remaining mint sprig. Serve with a straw.

Makes 1 mint julep.

I use less than a teaspoon of confectioners' sugar because I think bourbon tastes sweet anyway. If you are making mint juleps for a party, add the bourbon to the muddled mint mixture and pour into glasses filled with cracked ice. Make sure you have a mint sprig to garnish each glass. Don't forget the straw. And now is the time to use those cute cocktail napkins you have in that crowded kitchen drawer.

CHITLINS

In addition to bourbon and other traditionally southern-made drinks, the American South is home to several regional dishes, and one of them is chitlins.

Chitlins is southern for chitterlings, fried lower intestines from a pig. Like the extremities—ears, snouts, tails, and feet—they constituted the leftovers of hog butchering. These might sound inedible to some readers today, but for many people in American in the 1800s, these leftover bits and pieces of pigs were welcome on their empty plates. Affluent people lived "high on the hog" and ate the choicest cuts of pork (hence the origin of that saying). But the poorest folk, which in antebellum America always included the enslaved as well as the impoverished immigrant, had to make do with what they could get.

On the meager slave diet, chitlins were better than going hungry and at least added variety to the unchanging, limited rations of cornmeal and salt pork, both of which also provided only limited nutrients and vitamins. Available in the late fall—hog-butchering time because the air is getting colder—chitlins smelled repulsive and required very careful cleaning before cooking because of fecal bacteria lingering in the intestines.[13]

To make these more appetizing, slaves combined bits of African food traditions with the chitlins and other food available, either from their rations, small vegetable plots, or leftovers from the big house. Chitlins might be fried in pork fat over a quick, hot fire as a special treat, with collard or turnip greens stirred in. Or slaves might add okra, an African vegetable brought to the New World (and still prevalent in southern cuisine today), leftover corn bread, beans, herbs, water, and whatever else they could rustle up, and let it all simmer into a one-pot meal.

In the 1960s, black American culinary tradition donned the label "soul food" and included dishes like chitlins and collards along with southern fried chicken, pigs' feet, sweet potato pie, watermelon, and other tasty dishes with low-cost ingredients like Jell-O and macaroni and cheese.

I think chitlins are an acquired taste and not for the faint of heart. If you like pigs' ear and pigs' feet, chitlins might be for you. My friends who grew up eating chitlins like them very much. They remember being admonished as children to be careful whose chitlins they ate, because a careless cook might serve unhealthy chitlins.[14] Today, it's still a wise practice to know your food sources, whatever you are eating.

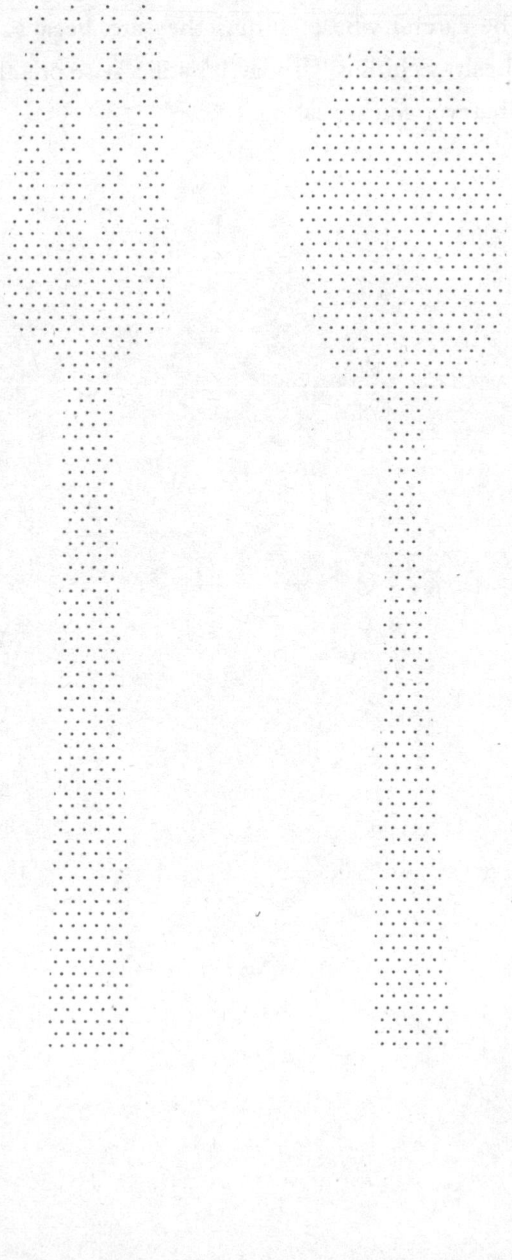

chapter 5

HARDTACK AND CHOP SUEY

From the Civil War to the Factory (1860–1875)

ar accelerates the rate of change, and the great watershed in American history known as the Civil War illustrates that truism. The war formally ended chattel slavery in the United States, boosted the development of a national economy, and hastened the rise of industrialism. During the war, Congress was able to enact a slate of progressive legislation, supported by President Lincoln, because the Southern states had withdrawn from the U.S. government to join the Confederacy.

No longer hog-tied by the South's opposition to these progressive reforms, the federal government successfully passed remarkable laws including the Homestead Act, the Morrell Act to create and support colleges, and the development of the Transcontinental Railroad. All of this legislation had an enormous impact on immigration, labor, education, agriculture, and ultimately the future of America. Today we focus so much attention on the heroism and tragedy of the battlefields that we often overlook the nonmilitary accomplishments of that era.

Take the Homestead Act, for example. Just imagine living in Europe and struggling to pay the tenant fees on the land you farm but will never own, because agricultural lands have been owned by the great noble families for generations. You hear about the passage of a new United States statute in 1862.

Under this groundbreaking law (no pun intended), in America, people could work hard for five years and then *own the land they farmed.* The government gave you a nice parcel of land—160 acres!—in return for your sweat equity and the improvements you made. This was a stunningly progressive concept. It's no wonder people picked up their lives, left the world they knew, and emigrated here.

Or the radical concept of a massive railroad linking both sides of the continent, a complex and futuristic vision that required thousands of workers. No wonder people immigrated here from all over the world. And they brought their culinary traditions with them.

The nature and availability of different foods reflected the realities of wartime, reconstruction, and the nation's unprecedented economic and social transformation. From the meager rations of the Confederate soldier to the Chinese American chop suey to a homesteader's rhubarb pie, the various dishes and diets that typify this era provide abundant examples of how this was a time of exceptional and fruitful change, not only in the American diet but in the very definition of what it meant to be American.

BITE 41

❧ LINCOLN'S ❧ FAVORITE CAKE

No one is more emblematic of the American Civil War era than Abraham Lincoln, the president who ranks number one (or second only to George Washington) in presidential popularity contests year after year. But as many history buffs know, in his day, Lincoln was not universally beloved. He was extremely controversial, and even his supporters feared he might not win reelection in 1864. Thankfully for the future of our nation, his admirers were stalwart, vocal, and articulate in their support.

Mary Todd Lincoln, however, would win no popularity contests as the president's complicated, demanding wife, either today or during her stay in the White House. She does, however, provide an interesting looking glass for examining women and their various roles in Civil War America. Mrs. Lincoln was not a typical female for her era, but her story and her recipe for White Almond Cake help us understand and perhaps sympathize with her a bit more.

The daughter of a prosperous merchant lawyer in Lexington, Kentucky, Mary grew up in a slave-holding family where the enslaved domestic servants alone outnumbered the large Todd family. Two of her brothers would later serve the Confederacy, leading some to question her loyalty. (Imagine if Eleanor Roosevelt's brother were an enemy officer during World War II!)

As the pampered youngest daughter, Mary had few household responsibilities while attending prestigious schools in Lexington. Fluent in French and passionate about reading, she was well educated for a female in nineteenth-century America. Mary also learned how to be a gracious hostess and a charming dance partner. But she had little experience in managing a household when she married Abraham Lincoln.

Accustomed to her family's well-trained house slaves in the South, she didn't cope well with the independent "hired girls" who helped out for a set fee in middle-class homes in Springfield and other communities across the North. Her lawyer husband traveled frequently for weeks at a time, following the circuit court around the state. Mary was often the single parent, left alone to manage their large house in the Illinois capital with three rambunctious boys.

Mary Todd Lincoln tried hard to fulfill the contemporary role for women during her era, which placed the married mother at the center of the domestic sphere. Popular magazines such as *Godey's Lady's Book*, to which Mary subscribed, emphasized the "cult of true womanhood" and the accomplishments of housewifery. As one commentator observed in *Ladies' Magazine*, "To render *home* happy is a woman's peculiar province; home is *her* world."[1]

Abraham Lincoln grew up in extremely rustic surroundings, with few clothes, a dirt floor, a tight food supply, and a harsh father. Eager to rise in the world, I think he found Mary's gentry background and family political connections appealing even if her emotional outbursts and excitability could be burdensome.

"My wife is as handsome as when she was a girl, and I, a poor nobody then, fell in love with her; and what's more, I have never fallen out," he wrote charmingly.[2] I doubt he minded if Mary was an indifferent housekeeper, although visitors made unkind observations about the state of her house in a day when genteel domesticity was the zenith of society's goal for women.

All of that said, she did learn to cook, and set a generous table. Mary brought a recipe for a white almond cake with her from Lexington and supposedly baked it for her tall fiancé when they were engaged in Springfield. Apparently, she also brought the cake recipe with her to the White House, where she was once again well equipped to plan menus and give instructions to the cook and kitchen staff, rather than keep up the household herself. Her husband ate with gusto terrapin stew, oysters, stuffed pheasant, and lobster pie as president. But he always retained his fondness for his wife's White Almond Cake.

MARY TODD LINCOLN'S
WHITE ALMOND CAKE

Thanks to Donna McCreary for sharing this recipe from her book Lincoln's Table.[3]

INGREDIENTS

► 1 cup blanched almonds
► 1 cup butter (2 sticks)
► 2 cups granulated sugar
► 3 cups all-purpose flour
► 3 teaspoons baking powder
► 1 cup whole milk or half-and-half
► 6 egg whites (best when eggs are at room temperature)
► 1 teaspoon almond extract

- ► 1 teaspoon vanilla extract
- ► Confectioners' sugar
- ► 2 tablespoons sliced almonds
- ► 1 to 2 scoops cherry ice cream

DIRECTIONS

Preheat oven to 350°F. Grease and flour two 9-inch layer-cake pans. Pulverize the almonds until they resemble coarse flour. Cream the butter and sugar with an electric beater or stand mixer until light yellow in color and fluffy. Sift flour and baking powder three times. Fold flour mix into creamed butter and sugar, alternating with milk, until well blended. Stir in almonds and beat well.

In a separate bowl, beat egg whites until they have stiff, firm peaks. Fold egg whites gently into batter with a rubber spatula. Add almond and vanilla extract. Pour batter into prepared pans and bake for 25 minutes, or until a skewer inserted comes out clean.

Cool for at least 20 minutes before inverting, then allow to completely cool before serving. Sift confectioners' sugar on top of each cake, or frost with a hard white icing, and create a layer cake, decorating the top with sliced almonds. Serve alone or with cherry ice cream.

Makes about 8 slices.

Baking powder could be purchased in a shop by the 1850s, which made cake baking much easier. A food processor works best for pulverizing the nuts. Make sure beaters are washed and dried thoroughly before beating the egg whites, or the whites will not stiffen properly. The egg yolks are not used in this recipe, so save them for another purpose.

BITE 42

❧ SOLDIERS' RATIONS ❧

North versus South

When the Civil War broke out in 1861, the Union and the Confederacy struggled painfully to organize their army supplies. Both sides initially employed the same list of rations presented in William Hardee's manual, *Rifle and Light Infantry Tactics*, published in 1855. This is hardly surprising, since the military leadership on both sides had trained together prior to the outbreak of the Civil War and, in many cases, fought together during the Mexican War as young men in the 1840s.

> *At night, a reckless Yank might creep out near enemy lines to trade coffee beans for tobacco with his foe during the Civil War.*

Despite the South's initial success on the battlefield, it reduced soldiers' rations by the end of the first year of the war and struggled from then on to feed its troops, even when they were in camp. The Confederate forces were already short of food.

In contrast, the Union army had perhaps the best-fed troops in the history of warfare. Food supplies for Yankee troops improved remarkably after a wasteful and somewhat corrupt beginning. (Unethical suppliers plagued the military with substandard and even rotten provisions on both sides of this war, the most deadly in U.S. history.) To help feed the Northern army, young soldiers, some just boys, herded droves of cattle and swine that had "escaped" from Southern plantations behind the regiments, providing a regular supply of fresh meat.

Engineers constructed huge bakeries at key ports and towns, enabling enlisted men to enjoy fresh bread when they were near supply depots. Thanks to a growing number of railroads and the eagle-eyed quartermaster general, Montgomery Meigs, the Yankees boasted an efficient production and distribution system that grew as the war dragged on. By the end of the war, bakers for Grant's Army of the Potomac at City Point, Virginia, were producing one hundred thousand rations of wheat bread a day for the soldiers.[4]

Of course, when they were on a long march, it was a different story. Union soldiers might live on hardtack, salt pork, beans, and coffee. Coffee was warming, stimulating, and readily available, and officers found it a far better drink for enlisted men than the tot of rum or whiskey that their forefathers consumed ninety years earlier in the Revolutionary War.

The Confederate troops had tobacco, but as the war raged on, they had little else. Union provisions seemed like a banquet compared to the parched corn and water that the Southern soldiers subsisted on toward the end of the war. As Andrew Smith points out in *Starving the South*, the North actively sought to destroy the food crops and supply lines of their enemy as it became clear that starvation was one of the ways to ensure eventual surrender. By the time of their surrender, Robert E. Lee's men weren't just hungry, they were starving. Meanwhile, the North enjoyed bountiful harvests, a boom in agricultural mechanization, and the ensuing increase in production.

That said, once Yankee regiments outdistanced their supply lines in the

The two Union soldiers in the front of this photograph are holding fresh loaves of bread to eat. It's commonly assumed the Civil War soldiers had to survive off hardtack and other minimal rations, but the Union army was actually the best fed in the world, especially when troops were close to their base camps.

Another factor that helped keep the Union soldier well fed was that the battle-field never extended north of Gettysburg, Pennsylvania. Thus no soldiers (Union or Confederate) trampled through New York wheat fields or drove live-stock away from New England pastures. The farmlands of the North stayed intact.

West and Deep South, they faced food shortages and relied on foraging. Hardtack, the tough, dry biscuit that epitomized army rations, followed them wherever they went. *Hardtack and Coffee*, a firsthand account of army life by Massachusetts veteran John D. Billings, provides descriptions of several ways soldiers managed to eat hardtack, which could be moldy, maggoty, or weevil-laden.

In the latter case, Billings recommended putting the hardtack in a hot cup of coffee. The weevils, drowning, would rise to the surface, where they could easily be skimmed out of the cup. And the hot coffee would soften the wooden texture of the wheat flour biscuit. Although America was primarily a tea-drinking nation in the antebellum period, the Northern soldiers' reliable supply of coffee—even on the march—may have turned the tide in favor of the stronger stimulant later in peacetime.[5]

BITE 43

FRIED CATFISH

The Southern Home Front

Broadly speaking, the soldiers' diet reflected the availability of food on the home front. Of course, individual civilian experience varied widely. But generally, households in the North ate better than their equivalents below the Mason-Dixon Line. Like their gray-clad soldiers, Confederate families faced a meager diet, sometimes supplemented with game. Yankees effectively cut off the salt supply through the embargo, so preserving meat was a challenge for Southern quartermasters and housewives alike.

Due to its reliance on cash crops, the South imported food from the

North during much of the antebellum period. So once war broke out, the South suffered a deficit of many kitchen basics for cooking. Plus, labor was in short supply, with the able-bodied white men serving in the military and the slaves leaving their owners in search of freedom. Crops rotted in the fields with no hands—black or white—to harvest them. Adding to the Southern woes, the weather in many areas was particularly punishing to Dixie farmers between 1862 and 1865, in stark contrast to the beneficent conditions the North experienced during those years.[6]

The siege of Vicksburg in 1863, when civilians ate rats to survive, presents the most famous tale of Southern civilian hardship. But hunger haunted much of the South. The Confederacy stood for decentralized government and never developed an administrative branch that could manage food distribution on a basic level. Extreme food shortages resulted in hunger and eventually bread riots, led primarily by women in Southern towns and cities whose menfolk were in the army. From Mobile, Alabama, to Richmond, Virginia, mothers and wives took to the streets to protest the lack of food and the high prices of bread, breaking windows and looting stores.[7]

Many white women learned how to cook for their families with ingredients they previously would have thrown out or given to their slaves. Recipes from this era frankly acknowledge the limitations on staples and groceries. For instance, roasted acorns served as a substitute for coffee beans, while a cracker-based filling produced a mock apple pie. Without the salt needed for meat preservation, one contemporary recipe for curing pork recommended hanging thin pieces of the raw meat against a southern wall so it would get more sun and dry quicker. The anonymous author included no advice on how to keep flies or maggots at bay.

When the Union army occupied an area, it used food as a way to gain civilian cooperation and support. Food became a key ingredient in the winning of the war.

Even after the Yankee army swept through a Rebel territory, there were sometimes squirrels in the trees, rabbits in the fields, and catfish or trout in the rivers that might be caught for dinner by civilians. Game and fish provided important sources of protein for all races and walks of life in war-ravaged regions. Catfish, not necessarily gourmet provender, is particularly associated with the South. Today it is often farm raised and quite small, but the big old catfish feeding on the bottom of lakes and rivers long ago could make a meal for a whole family and maybe some friends too.

FRIED CATFISH

This very simple recipe would have been a welcomed main dish on any plate during the Civil War, particularly in the South. Salt and pepper were luxuries on the Southern home front, although cornmeal was available. A platter of steaming white rice and stewed tomatoes served alongside the fresh catfish would make this a feast during the war years.

INGREDIENTS

- ► 4 tablespoons lard
- ► 2 ½-pound catfish fillets
- ► 1 cup flour or cornmeal
- ► 1 pinch thyme
- ► Salt and freshly ground black pepper to taste

DIRECTIONS

Melt the lard in a heavy skillet over medium-high heat. Roll the catfish in cornmeal with thyme added, shake off excess, and fry in the lard. Cook on each side for 4 minutes and season with salt and pepper.

Serves 4 as a main dish.

Serve with white rice and 2 cups stewed tomatoes to which 1 tablespoon vinegar, 1 teaspoon brown sugar, and a pinch of cayenne pepper have been added.

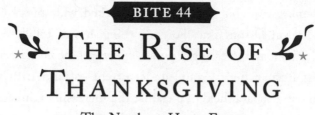

THE RISE OF THANKSGIVING

The Northern Home Front

We learned a little bit about the origins of this truly American holiday in Bite 20 about turkey, but now it's time to circle back and finish where the story left off. In November 1860, on the eve of the Civil War, many Americans in the North sat down to a Thanksgiving feast, gathering family around a table piled high with food and drink. In New England and the Midwest, a region primarily settled by Yankees, Thanksgiving was widely celebrated as one of the most important holidays of the year.

As there was no nationally recognized date established for the observance yet, the day was set by each governor, and it generally fell on a Thursday in November for good reason. Late fall brought the cold weather that hog and poultry slaughtering required on the farms, and with the harvest complete, households often had the time needed for the extra preserving and baking and brewing expected for a true day of feasting.

By the mid-nineteenth century, the harvest festival held by early English settlers with their American Indian guests in Plymouth, Massachusetts, in 1621 provided an iconic combination of piety, legend, and patriotism that flourished particularly in the North. *Godey's Lady's Book* and other popular magazines featured recipes and editorials that embraced the heart and hearth of this agrarian holiday that was uniquely American but had roots in ancient times.

Sarah Josepha Hale, longtime "editress" of *Godey's*, campaigned for many years to have Thanksgiving declared a national holiday. She saw it as a unifying observance that could override the rising influence of sectionalism pervading the nation. "Thanksgiving should be a national festival observed by all the people as an exponent of our national virtues," Hale wrote in 1846, a theory she advanced for the next seventeen years, eventually with success.[8]

Individual Southern families, especially those with Yankee relatives, enjoyed Thanksgiving, but as a holiday it lacked widespread observance below the Mason-Dixon Line. For one thing, it was totally eclipsed by Christmas, a holiday increasingly popular in the nineteenth century. But reflecting its Puritan heritage and undercurrents of anti-Irish sentiment, Protestant New England held out against the increasing Christmas observance, sensing lingering paganism—or perhaps even worse in their eyes, "popery"—in the revels of late December. The Massachusetts legislature somewhat begrudgingly accorded Christmas legal status as a holiday in 1856, and while more and more families adopted the "new" tradition, it was Thanksgiving that remained central to New England heritage—and in many ways, still does today.

Meanwhile, some Southern leaders increasingly saw Thanksgiving as a Yankee holiday laced with overtones of antislavery, due in part to the practice of giving public speeches on that day. Politicians in New England observed the holiday by speaking at length on patriotic topics with themes of freedom and liberty. Newspapers would print and circulate the remarks, some of which referenced the evils of slavery. Henry Wise, governor of Virginia from 1856 to 1860 and later a brigadier general in the Confederate army, pointedly refused to recognize Thanksgiving as a state holiday, claiming the observance supported "other causes" like abolitionism.

Today, in hindsight, the tide of war was turning when Lincoln made Thanksgiving a national holiday, so it may seem like a frivolous political act to us. But at that point there was no guarantee of Northern victory. Lincoln may have wanted to reap some political gain from this relatively happy moment.

In October 1863, three months after the Union victories in Gettysburg and Vicksburg, President Lincoln proclaimed the first national Thanksgiving to be celebrated as a feast day. That month also witnessed an especially fruitful harvest throughout most of the Union.

Sarah Josepha Hale's campaign had finally borne fruit—she had written to the president personally, asking him to declare a national Thanksgiving on the fourth Thursday of November. The timing of her request was perfect, whether she realized it or not. Lincoln's proclamation acknowledged the bountiful harvests, robust economy, and alluded to the military victories, giving him the opportunity to let the Union bask for a moment in "the providence of Almighty God."[9]

The following year, prosperous New Yorker George W. Blount led a civilian campaign to bring Thanksgiving to every Union soldier and sailor fighting in Virginia. Local newspapers and charities embraced the call, while shipping agents, produce sellers, and packers pledged their support. Soon the goal blossomed into reaching as many Yankee soldiers and sailors as possible. The North witnessed an enormous outpouring of bounty.

In *Starving the South*, Smith describes this unprecedented effort, such as the "Soldiers Aid" of Norwich, Connecticut, which organized the shipment of 215 turkeys, 199 pies, and a long list of other comestibles to Connecticut regiments in the Armies of the James and the Shenandoah. Ladies' organizations across the Union cooked sumptuous Thanksgiving feasts for their state army posts and veterans' hospitals. Yankee supply steamers delivered barrels of food to sailors along the coast and up the rivers of Dixie.

> *To give a sense of the immensity of the civilian campaign to bring Thanksgiving to Union soldiers everywhere, New Yorkers alone sent more than three hundred thousand pounds of poultry with literally tons of other festive ingredients to the Army of the Potomac.*

Judging by the letters and journals of the appreciative Yankee troops, the Thanksgiving of 1864 was an immensely popular success, and most of the shipments reached their intended audience. When Sherman's army reached Savannah on December 10, the Union supply ships furnished the soldiers with the Thanksgiving dinner they had missed during their march to the sea.[10]

Food is enormously evocative of home at all times, but particularly for soldiers far away from their families. Federal military leaders understood just how important a Thanksgiving meal would be for morale of the enlisted men. During the conflicts that have followed the Civil War, civilian and military leaders have worked together to guarantee that American men and women serving far from home get a plate of turkey with all the trimmings on or close to Thanksgiving.

RAILROAD WORKERS AND CHOP SUEY

Most Americans remember Abraham Lincoln as a wartime president and the Great Emancipator. He was both, plus a big supporter of the latest technology. As the sixteenth president, he understood the importance and impact of the telegraph—the first fast, long-distance form of modern communication—long before many of his peers.

Lincoln also supported the seemingly crazy scheme for the Transcontinental Railroad to link the Pacific and the Atlantic coasts. Completed in 1869, just four years after his assassination, this was the longest railroad in the world at that time. It created the possibility of national markets for food and beverages that had once been strictly local or regional. And members of the railroad workforce introduced Chinese food to America.

The construction of the railroad started at two different origins: from Omaha, Nebraska, heading west and from Sacramento heading east. Both segments were massive undertakings, but the one originating in California faced the biggest uphill battle. Due to the scale and elevation of the charted path, the engineers realized they would need a lot of explosives to open up the rock face of the Sierra Nevada range. And explosives in the 1860s were dangerously unpredictable, costing workmen's lives. In 1865, the Central Pacific Railroad needed workmen, and they needed gangs who understood how to set explosives. So the railroad owners did something they had sworn not to do—they hired immigrant Chinese workers.

Prior to the Civil War, Chinese peasants began to emigrate from Guangdong (Canton) province in southern China to California to escape the poverty and overpopulation of their homeland. They sought their fortunes in the gold fields, prospecting, working as day laborers, and serving as domestics. Most of the Chinese immigrants were male and planned to return to their families one day. In the meantime they worked hard and faced tremendous prejudice from the white population, particularly from other new arrivals like the Irish.

When Irish railroad workers threatened to quit rather than work alongside the "heathen Chinee" or "Celestials," the Central Pacific refused to be intimidated. After all, the Chinese workers were diligent, rarely drank, and could be paid around 20 to 30 percent less than the white workers. The Chinese also knew how to handle dynamite. Plus, the railroad had no obligation to provide their Asian workers with board—the free meals they provided their non-Asian laborers. In other words, the Chinese could be ruthlessly exploited, and the Central Pacific labor contractor, Charles Crocker, and foreman, James Strobridge, did exactly that during and after the Civil War.[11]

THE IRONY OF EXPLOITATION

Not receiving food from railroad management proved to be a real boon for the Chinese. While the white workers received unhealthy rations of salt or fresh beef and potatoes, plus a ration of whiskey or rum from time to time, and drank water from possibly tainted sources, the Chinese carried food purchased from Asian merchants in San Francisco and cooked their own meals, primarily dried seafood, greens, and rice.

They also kept live chickens for their eggs and occasional fresh meat. An important key to their health was drinking only boiled water, usually in the form of lukewarm tea during the day. Thus the Chinese did not suffer from the dysentery and malnourishment of the workers who received poor quality food as part of their pay. They also washed their clothes and bathed regularly, a suspicious habit frowned upon by some of the laborers who had immigrated from western Europe.

Once the railroad was completed, some Chinese workers went on to help build railroads in different regions. Others settled in western towns or returned to San Francisco and found new jobs. And a few opened small restaurants, which started to introduce the American palate to exotic Asian cuisine, slowly but surely. Along with several authentic dishes, Chinese restaurants sometimes served a dish known as chop suey which became popular with American customers.

The name "chop suey" is the anglicized version of Cantonese words meaning "cooked entrails" or "odds and ends" or even "leftovers." (In Mandarin it is pronounced "cha sui" with no *p*.) It is a mixture of meat and eggs, quickly stir-fried with vegetables in a sauce and served over rice. If you substitute noodles for the rice you have chow mein—odds and ends over noodles. There are lots of stories about how chow mein and chop suey were invented, but no real evidence. However, we do know that, by the 1880s, Americans recognized chop suey and chow mein as Chinese food they liked. The fact that no restaurant in China served either dish worried no one. This was, and still is, uniquely American Chinese food.[12]

BITE 46

❧ BORDEN'S CANNED CONDENSED MILK

Railroads were an enormous force for change in American society and its food traditions. Even regional trains expanded the distribution of nonperishable items, creating huge fortunes and brand names that still line the shelves of our grocery stores. A classic example of a food entrepreneur who benefited from the transportation revolution of the nineteenth century was Gail Borden.

In 1857, Gail Borden Jr. founded a food company based on new technology, creating one of the first engineered foods in American history. (As we'll see, it would not be the last!) After a series of trials and errors, Borden perfected the process by which cow's milk had most of the water removed and some sugar added. The sugar, combined with the heated canning process, gave condensed milk a shelf life of many years, something that was a revolutionary invention in this largely prerefrigeration era. Borden became one of the first national food suppliers and changed the way people drank, baked, and ate.

Today, few Americans buy condensed milk unless they are making a specialty dessert, like pumpkin pie or dulce de leche. There was a time, however, when canned condensed milk was a newfangled luxury product—and the cornerstone of an American enterprise.

So how did he create this once-miraculous invention? In 1851, Borden returned home from a trip to Britain, troubled by the illness and death of child passengers on board ship. Apparently, they had been given milk from sickly cows kept on board. Refrigeration was so limited in those days that, in warm weather, you had to be near healthy cows to get drinkable, fresh milk. So Borden was determined to figure out a way to preserve milk that would last a long time without refrigeration.

After a series of failures and with the help of an investor, Jeremiah Milbank, Borden perfected the process for condensing milk. He contracted

An advertisement for Borden's Condensed Milk depicting how babies "love it."

with numerous local dairies, requiring his suppliers to observe strict sanitary standards in the milking process. Sold as Eagle Brand, with a bald eagle spreading its wings across the bright label to show the company's patriotism and American roots, Borden's condensed product developed a reputation as healthy, pure, and useful milk.

Borden's timing was perfect. With a modern factory, surrounded by eager suppliers and close to transportation to the northeastern urban markets, his business grew. Incidents of city children being poisoned by bad milk encouraged mothers to buy the canned product and protect their sons and daughters (as well as themselves).

On the march, lucky officers bought Borden's milk from "sutlers" (peddlers who sold goods to personnel) to lighten their coffee, but in camp, the Yankee foot soldier and sailor enjoyed it as well.

Then, with the outbreak of the Civil War, his business really boomed. The federal government bought up quantities for the military's use. Condensed milk was used specifically as a field ration, since it supplied 1,300 calories in a ten-ounce can and seemed to last forever, regardless of weather. Two hundred dairy farmers supplied twenty thousand gallons of fresh milk a day to Borden to fulfill his lucrative War Department contract.

When the war ended and the troops returned home, they created an ongoing market demand for Borden's condensed milk. The railroads that now crisscrossed the North (and would link the coasts by 1869) transported the cans across the nation. Following the trajectory of a burgeoning U.S. economy, Borden's dairy enterprise grew into one of the first national brands.[13]

BITE 47

❧ BEER AND PRETZELS ❧

The end of the Civil War brought cultural changes, and when there are cultural changes, changes in food and drink are not far behind, because they are so closely linked. For example, as the number of German immigrants rose, many aspects of German life—from their university system to their gymnasiums to their food traditions—became an intrinsic part of American culture. German lager-style beer and pretzels grew in popularity, and the beer hall was close behind.

Germans made up the largest single immigrant group in nineteenth-century United States. Like the Irish, they arrived in families and as individual workers. By the mid-1800s the Germans—Catholic, Protestant, and, in lesser numbers, Jews—were settling everywhere, and their religious denomination would determine the other ethnic groups with whom they intermarried. (Today, many Irish, Italian, and English families are actually partly or mostly German, a fact that was sometimes "forgotten" after World Wars I and II.)[14]

The Germans brought recipes and techniques for two things that today seem all-American: beer and pretzels. We know that the early colonists drank beer, but if you recall, it was sometimes murky and unpredictable due to the erratic character of the wild yeast that occurs naturally in North America, as well as having a propensity to turn sour, especially in the warm weather. The Germans brought a different kind of brewing that recreated the European lager-style beer, along with a commitment to careful excellence in the creation of that drink.

Its taste was consistent and delicious compared to previous American-made beers and quickly caught on. Beer halls serving tankards of foamy lager—with plates of sausage, potatoes, sauerkraut, and dark bread—became popular in the big cities, particularly places like New York, Philadelphia, St. Louis, and Chicago. Unlike the male-only environment of American bars and saloons, most beer halls welcomed families.

Along with beer, the German immigrants brought their pretzel recipes. Pretzels have been around for a long time, and many European countries had their own regional pretzel recipes by the first millennium. In Germany, families ate pretzels during Lent, at wedding celebrations, where the knotted dough symbolized the binds of marriage, and at any other time throughout the year. They made crispy thin pretzels and chewy fat pretzels. When Germans immigrated to the United States, pretzel baking traveled along with their other cultural baggage. Today, 80 percent of all pretzels made in America are still produced by commercial bakeries with German roots.

Many of the old beer-brewing families—such as Pabst, Stroh, Anheuser, Busch, and Coors—share those German roots as well. In 1860, the year before the outbreak of the Civil War, Eberhard Anheuser purchased a failing brewery in St. Louis, which he named the Anheuser Company. His young daughter Lilly married a brewery supplier, Adolphus Busch, who took over the family business when his father-in-law passed away a few years later, and added his name to the company's title. Renamed the Anheuser-Busch Company, the brewery invested in marketing, technology, and new product development. By the 1870s, with the invention of refrigerated railroad cars, Adolphus Busch realized he could ship his product nationally. (An inventor in Baltimore developed the crimped metal "crown cap," still used today, that enabled manufacturers in the 1890s to ship beer and soda in bottles without them exploding.) Recognizing that a national product required national marketing, the Anheuser-Busch Company bought advertising throughout the United States at a time when most breweries were strictly local.

Three German Americans enjoy mugs of beer in the 1800s.

Busch had a new product he wanted everyone to know about: a bottled beer called Budweiser. His timing was perfect. Budweiser became the first national brand of beer, and Anheuser-Busch eventually grew into the largest brewery in the world.

Today, local beers are making a comeback. "Think globally, drink locally" is the new rallying cry among many dedicated beer drinkers, and microbreweries often sponsor Oktoberfests in their communities. But there was a time when a national beer in a branded bottle, clean and predictable, represented an exciting new advance. Today, whenever and wherever people drink beer, they are taking a sip of history.

OKTOBERFEST

Oktoberfest, an annual celebration of beer, food, and music, began in 1810 in Munich, Germany, as the public toasted the marriage of their Crown Prince Ludwig with Princess Therese of Sachsen-Hildburghausen, whose name is too good to resist including here. Everyone had so much fun that they decided to hold the gathering again in 1811, and it grew into a yearly event. Today, approximately six million people from around the world visit Munich in late September or early October to take part in the beer-drinking festivities.

Other countries, from Brazil to India, started observing Oktoberfest largely toward the end of the twentieth century. La Crosse, Wisconsin, sponsored one of the first Oktoberfests in America in 1961. Since then, literally hundreds of communities around the country have started hosting their own Oktoberfest in midfall. As Prince Ludwig would say, "Prost!"

GERMAN BEER HALL FOOD

It is easy to replicate dining at a nineteenth-century beer hall in terms of the food and drink, but I would avoid the messiness of sprinkling traditional sawdust on the floor to protect it, if I were you. In warm weather, some beer halls would expand to picnic tables outside and offer a beer garden. The traditional menu does provide great food for an evening outdoors with family and friends by the grill. If you're interested in throwing a beer hall–style party for Oktoberfest or any occasion, here's a suggested menu.

BEER HALL MENU FOR OKTOBERFEST AT HOME

Beverages:
- German beer—all kinds.
- Lemonade that has literally been watered down by 50 percent.

Food:
- Pretzels, fat and thin. (Avoid flavored pretzels, if possible, to preserve authenticity.)
- German sausages: knackwurst, bratwurst, weisswurst, and frankfurters are good to try. See what you can find. You should offer at least two kinds of sausage. (Note: This is not the time to try Italian sausage if you want to preserve authenticity.)
- Sauerkraut, served hot.
- Radishes, steamed carrots, celery sticks, and pickled beets, attractively arranged on a platter lined with lettuce leaves.
- Sliced boiled potatoes, served at room temperature with a grainy mustard vinaigrette and sprinkled with parsley, also on a platter.
- Cucumber salad with dill.
- Assorted breads and rolls, including rye, offered with butter and a variety of mustards.

Dessert:

- A traditional German dessert is apple strudel. If you use frozen puff pastry, this is not hard to make, and there are many good recipes online. Or buy one from a good local bakery. Serve with homemade shlag: 1 cup heavy whipped cream to which you have added ¼ cup sugar, 1 teaspoon vanilla, and 2 tablespoons apple brandy or schnapps.

BITE 48
❧ PASTA WITH ❧
RED SAUCE

The Germans did not invent beer in America, but their brewing methods increased its popularity throughout the country after the Civil War. The history of pasta in America follows a similar trajectory. We know that Americans ate macaroni by 1800. However, the later waves of Italian immigrants brought a flotilla of pasta recipes with them across the Atlantic and eventually made pasta widely popular in this country.

The Chinese ate noodles more than four thousand years ago, and the ancient Etruscans, the Romans' precursors in Italy, produced pasta as well around 500 BC. The Muslims helped spread the enjoyment of pasta as their empire expanded in the 700s. During the great Age of Exploration, ships' chandlers sold noodles to the cooks, who appreciated the fact that dried pasta had a long shelf life. Contrary to a popular myth, famed explorer Marco Polo did not introduce pasta to Italy, although he did record people eating noodles in China.

So how does pasta overall figure into America's culinary past? It started with the type of pasta that's the costar in the quintessential American dish, macaroni and cheese. We know that Thomas Jefferson enjoyed macaroni in Paris, and served it at Monticello and the White House. Recipes for baked macaroni appeared in popular cookbooks in antebellum America, but other pasta like spaghetti remained a slightly obscure and foreign dish for many people.

Ironically, in 1848 a French immigrant, not an Italian, established the first pasta factory in America. Antoine Zerega opened his horse-powered factory on the Brooklyn waterfront and dried the strands of pasta on the roof. His sons inherited the business. Today, their heirs still run a successful pasta business aimed primarily at restaurants and other suppliers, producing more than two hundred million pounds of dried pasta a year.[15]

The first recipe for pasta with tomato sauce appeared in an Italian cookbook by Neapolitan chef Francesco Leonardi in 1791. It took another fifty-six years for tomato sauce to show up in an American cookbook, although people may have enjoyed it for several years before the recipe was published. Macaroni and cheese beat it by a few decades; it was already a popular dish by the 1830s.

Victorian Americans would be startled by the refined reputation of the gourmet pasta dishes served in elegant Italian restaurants today. Americans of northern European descent sometimes sneered at spaghetti with red sauce, due to deeply rooted class prejudice against Italian immigrants.

Italians occupied some of the lower rungs of the social ladder in the late 1880s, and while this would change as the twentieth century progressed, they encountered harsh prejudice in their new home. One of the largest mass lynchings in American history took place in New Orleans in 1891, when eleven Italians were hanged by a mob after nineteen of them had been acquitted of murdering the local police chief. Two had nothing to do with the case, but were dragged out of jail by the mob at the same time and hanged despite their lack of involvement.[16]

The Irish, now increasing acculturated, joined the WASPs by wrinkling their noses in disgust at the pungent smell of garlic cooked in olive oil. But the Italians prioritized their food like they prioritized their family, and eventually Italian cooking

HUNGRY ITALIANS AND SPAGHETTI AND MEATBALLS

About four million Italians, mostly from southern Italy, immigrated to the United States between 1880 and 1920. Economic conditions had worsened for southern Italians after the unification of Italy in 1861, and they fled the heavy taxes and lack of economic opportunities. They also faced real hunger in their home country.

Many Italians arrived with little education. The men worked primarily as manual laborers, particularly in urban public works, while the married women often worked on sewing projects at home. When they could raise the capital, some Italian Americans also started small truck farms in New Jersey and on Long Island, selling fruits and vegetables to small groceries and street peddlers in New York City. The comparatively low price of food, especially meat, delighted the new immigrants, who wrote to their relatives in the old country about America's bounty.

Even a generation later, the abundance of food surprised Italian arrivals. A miner's wife, Rosa Cavalleri, remembered telling her family about her life in the new country when she returned to bring back her young son, whom she had left in southern Italy with his grandparents. "Mamma mia, but that was hard [for them] to believe—poor people in America eating meat every day!"[17] The availability of meat in the United States would change traditional Italian recipes and introduce a new dish, spaghetti with meatballs, that was born in the New World.

became extraordinarily popular. It was just foreign enough to be exotic, but not so unidentifiable as to be off-putting to the less adventurous American.

By the 1900s, versions of Italian food became mass-produced. In 1887, Alphonse Biardot and his two sons started a food business, producing canned pasta in sauce, along with soups and gravies. The Biardot brand, Franco-American, was purchased by the Campbell Soup Company in 1915 and presented a variety of canned ersatz pasta dishes for generations until the 1990s.

Franco-American SpaghettiOs delighted kids and famished babysitters (myself included) for many years and still is popular today. The introduction of factory-produced tomato sauce won the hearts of consumers as well. Today, the Ragu brand of spaghetti sauce, first launched in 1937 by the Contisano family in Rochester, New York, is the most popular in the United States and Europe. It is owned by Unilever, a multinational conglomerate.

Until the mid-twentieth century, pasta with tomato sauce was simply an easy, cheap dinner, considered more of a blue-collar staple than an example of delicious dining. Today, a shopper may find excellent fresh or dried pasta, in a variety of shapes and colors, in restaurants, stores, and online. Diners of all ages and sophistication can find a fresh, canned, or frozen tomato sauce—from a meaty Bolognese to a vegan version—that suits their fancy, or easily produce their own. And even old-line WASPs swoon in delight at the delicious smells emanating from an Italian American kitchen.

BITE 49

LUNCH PAILS

The Italians were just one of many European peoples who immigrated to the United States in search of opportunity, bringing their food culture with them. With the rise of heavy industry, the need for laborers increased. And the demand for the coal that powered the railroads and the factories increased as well.

The rich anthracite coalfields of Pennsylvania provided the fuel for the steel industry, one of the backbones of the American economy for the century following the Civil War. The hard, bright anthracite burned more cleanly and at a higher heat than bituminous coal. But it required a skilled miner to harvest it.

Immigrants poured into the United States when the Civil War ended, looking for land and work in what appeared to be an ever-expanding economy. The coalfields near Scranton, Pennsylvania, attracted experienced miners from southern Wales, on Britain's southwest coast. Once an independent Celtic nation, Wales had its own language and cultural traditions. Absorbed by England in 1543, the Welsh were often considered second-class citizens well into the 1900s by the other members of Great Britain.

But in the United States, the Welsh fit right in. They were Protestant, they spoke English, and they brought their mining skills with them. Some of them ended up managing the mines, rising through the ranks due to their own hard work and acumen, and assimilating quite easily compared to other European ethnic groups. Welsh food also paralleled the meat and dairy-rich English diet, with potatoes, carrots, and cabbage as frequent additions. Their love of leeks is particularly noteworthy, and many Welsh recipes include this tall green member of the onion family.

People have cultivated leeks for at least three thousand years in the Middle East and Mediterranean region. The Bible records in the Book of Numbers that the ancient Hebrews ate leeks. Easily grown in sandy soil, and milder and sweeter than most onions, leeks were popular with the Romans and spread throughout Europe and the British Isles. The Welsh grew leeks by the seventh century AD when, according to legend, King Cadwaladr (or Cadwallader) and his troops wore leeks in their hats to differentiate themselves from their enemies on the battlefield. Fast-forward about 1,300 years, and the Welsh still grew leeks, which were also popular in America.

Leeks might be served at any meal, eaten at home around a table or at work as part of lunch. A miner carried his lunch in a tightly covered three-level pail. The bottom of the bucket contained hot tea, which he would drink and which also kept his food warm. The second layer held his food, perhaps ham with bread or, if he were a lucky Welshman, a leek-and-cheese pasty. Either one would be wrapped in oiled brown paper or a small cloth to keep it fresh.

The top level of the lunch bucket held fruit or a sweet biscuit if the miner's wife could spare it. You can still find miners' lunch buckets (or pails) at flea markets, online, or at antique stores especially in coal-mining districts like West Virginia and Kentucky. Often they are impregnated with black coal dust, tragically like the lungs of the old miners.[18]

FROM LUNCH PAILS TO LUNCH BOXES

Metal lunch pails represent a marked change in the work life of Americans, whatever their national heritage. In the traditional agricultural rhythms of life, workers ate meals at home, or food was transported to the fields in baskets, often by young family members. Bringing a lunch pail to the mine or the factory symbolizes the rise of the industrial age, when people worked outside the family-based workshop or the farm. The men and women who carried lunch pails also bore a class indicator, as a worker who did not sit down for the luxury of a hot meal at midday.

Children carried wrapped food in leather or muslin pouches or baskets to school from colonial days into the 1800s, until popular items like crackers and tobacco were commonly sold in tin boxes. The tin boxes could be reused as storage containers. By the 1880s, parents began converting these tins into children's lunch boxes.

The first retail lunch boxes specifically for kids appeared on the market in 1902, sporting noncommercial images of cheerful children at play. But that absence of a marketing presence did not last for long. Walt Disney pioneered the use of entertainment stars on lunch boxes in 1935, when his Mickey Mouse character appeared on one design and was embraced by children and their parents who could afford the lunch boxes.[19]

The baby boomer generation, born after World War II, grew up in the heyday of colorful, square lunch boxes emblazoned with TV and Hollywood heroes and heroines. In 1953, a children's lunch-box design featuring singing cowboy Roy Rogers and cowgirl Dale Evans was a huge hit: 2.5 million of this one design were sold that year. Meanwhile, working men and women carried rectangular lunch boxes with domed lids that could hold a thermos of hot coffee and their midday meal.[20]

In the twenty-first century, more children and adults buy their lunch at noontime than ever before, and lunch-box sales have suffered. But bringing one's own food guarantees the right ingredients and, in many cases, significant savings. For a lot of Americans, it's the smart choice.

BITE 50

❧ RHUBARB ☙

A Seasonal Food

Today, with the reemerging locavore movements, people are entranced by consuming food that is grown in their own area and in season, or seasonal. They are trying to eat in a more environmentally friendly way, rather than expecting to be able to consume all types of produce at any time of year.

Of course, before the rise of commercial canning in the late nineteenth century and the later development of frozen food, many types of fruits and vegetables could only be consumed fresh during a limited season or dried (which reduced some of the flavor and nutrients). Eating certain foods out of season was either a fabulous luxury or not such a good idea, as the food would not be tasty and might even be bad. Farmwives spent significant energy trying to preserve food past its season, by drying or carefully storing apples and root vegetables in root cellars, where they would grow limp and wizened but still edible.

Most fresh foods would star at meals for a few weeks and then disappear until they returned the following year to be greeted with enthusiasm and large appetites. Today, most Americans' encounters with a seasonal food are primarily limited to the summer-only availability of locally grown corn on the cob and tomatoes at farm stands and markets. But one other food is still exceedingly hard to find out of season: rhubarb, a welcome harbinger of spring and summer.

Rhubarb is an old-fashioned plant that is easy to grow if you have a sunny space and like to water the garden. Originally rhubarb came from northern Asia, and the Chinese valued the root of the plant for more than three thousand years for its powerful laxative effects. Rhubarb root imported from the banks of the Volga commanded a high price in medieval and Renaissance Europe, when doctors routinely prescribed purges to treat various maladies, along with bloodlettings and emetics.

The reddish stalks, fibrous and crisply juicy like celery, grew popular as a food only when sugar became more available in the late 1600s. Rhubarb is so sour that many recipes require a high ratio of sugar to its cooked fruit to create palatable (and frankly delicious) jams, sauces, and baked goods.

The leaves are very acidic and considered poisonous, although you would have to eat about three pounds of the horrendously sour leaves for a toxic dose. So—all you mystery writers out there—forget death by rhubarb leaves.

> *Rhubarb was so valued in medieval and Renaissance Europe that merchants paid more for powdered rhubarb root than they did for cinnamon or opium.*

Northern Europeans brought rootstock with them to the New World, and farmers here grew it in a corner of their kitchen gardens or in a sunny space in their orchards. It traveled with them as they moved westward. A year after the rootstock was planted in the spring, the stalks were ready for harvest, and they would come back year after year. Sarah Josepha Hale, the editor of *Godey's Lady's Book*, suggested a name change, observing, "In England they call this 'spring fruit,' which is a much more relishing name than rhubarb."[21]

Other Americans, such as the novelist Laura Ingalls Wilder, referred to it as "pie plant," since the diced stalks, mixed with sugar and a little spice, produced the first fresh fruit filling for pies in the late spring.[22] I often think how happily homesteaders in the nineteenth century must have greeted their rhubarb. It was their first fresh fruit after a long winter. Traditionally, it combines well with strawberries, which also make their appearance early in the growing season.

Today, you can buy the rootstock online or at a nursery to grow your own rhubarb. In the early spring, an established plant cracks the soil with the force of the emerging leaves and stalks. It is quite impressive. Rhubarb can be harvested in a few weeks once it breaks through the ground, and as mentioned, it returns generously every year, if tended in a sunny spot with well-rotted manure and lots of water.

STRAWBERRY RHUBARB PIE

Americans ate pie at every meal in the nineteenth century. To be historically accurate, I heartily encourage you to do the same. You will need a double-crust pastry recipe for this pie, or just buy two premade pie-dough circles. I get the kind that is sold in rolls, so I can use my own pans. Another option is to create a kind of crumble topping by mashing together oats, sugar, and butter in a bowl with a pastry blender and sprinkling on top of the rhubarb mixture before popping it into the oven.

INGREDIENTS

- 2 pie-dough circles
- 2 tablespoons cornstarch
- ¼ cup orange juice or water
- 1 cup granulated sugar
- 1 teaspoon salt
- ½ teaspoon allspice
- ½ teaspoon ground cinnamon
- 2½ cups diced rhubarb stalk
- 2½ cups fresh strawberries, hulled and sliced
- 2 tablespoons butter, sliced into 6 pieces

DIRECTIONS

Preheat oven to 425°F. Line a 9-inch pie pan with pastry dough. Use pie weights (or dried beans), if you can, to keep the dough from puffing while it bakes. Bake the pie crust for 10 minutes. Remove from oven.

Dissolve the cornstarch in the orange juice or water until very smooth. Mix with the sugar, salt, allspice, and cinnamon until well blended. Combine the cut-up fruit with the sugar mixture in a large bowl. Let sit for 15 minutes.

Pour mixture into the pie shell, and dot with butter slices. Cover with the pie dough, and cut steam holes into the top pastry. Cover the edges of the pie crust with foil so they don't burn. Place the pie pan on a cookie sheet lined with parchment or aluminum foil, so your cookie sheet and oven are protected from the mess if your pie bubbles over. Reduce heat to 350°F once the pie is in the oven. Bake for 1 hour. Serve warm with ice cream or for breakfast with a great cup of coffee.

Serves 6 to 8.

Do not use any of the rhubarb leaves, which you now know are toxic. For the steam holes, try slicing in your kids' or a guest's initials—this is always a hit. Or weave a lattice top if you are feeling ambitious. Did you buy your pie crust? Then really, you should make the effort to try this. It looks super impressive, and it's not difficult if you plan it out and use a ruler or a careful eye. (Watch a how-to video online if you've never done this or witnessed it being done. Howdini has a good video on YouTube.)

chapter 6

BAKED ALASKA AND BARBECUE

The Gilded Age, the Gritty Age (1870–1900)

T he period between 1870 and 1900 witnessed the growth of profound contrasts in the distribution of wealth and standards of living. Through hard work, innovation, luck, or inheritance, or a combination of these, some Americans had grown rich in every generation since the days of the first settlements. Being rich per se was not new. But in the late nineteenth century, a relatively small group of industrialists and financiers created enormous private wealth, primarily through the rise of the business corporation. Commentators have pinned a variety of labels on men like John Jacob Astor, Andrew Carnegie, John D. Rockefeller, J. P. Morgan, and Cornelius Vanderbilt, ranging from robber barons to industrial statesmen. There is lots of room for different interpretations.

The presence of this dynastic wealth, however, increased the number of affluent businessmen and professionals who in turn had larger disposable incomes. And these incomes enabled the rise of fancy restaurants, like Delmonico's in New York, and elegant hotels, like the Palmer House in Chicago. For the first time in the United States, talented chefs developed celebrity status, due not only to their skills but also because enough people could afford elaborate, multi-course dinners served with fine wines in luxurious surroundings. Ultimately, the Gilded Age witnessed an increased sophistication in the American menu.

Immigrants and "native" laborers, working in the first large-scale industries, along with many small farmers, lived in stark contrast to the elite. As opposed to the Gilded Age, they lived in what I like to call "the Gritty Age." Toward the end of the century, bloody strikes at factories, mines, and railroads protested the inhumane working conditions that helped generate the high corporate profits.

Mark Twain coined the term *The Gilded Age* as the title of a satirical novel he coauthored with Charles Dudley Warner in 1872. To Twain, the era glittered with a thin layer of gold barely covering the dirty corruption that pervaded politics, business, and society as a whole.

Still, social mobility in America allowed for the continued growth of an aspirational middle class. The traditional, slow-paced rhythms of daily life quickened in response to new inventions, like the telephone and the bicycle, and the accessibility of railroad and streetcar transportation. And the introduction of new foods, both gilded and gritty, would significantly change the way Americans ate.

BITE 51
❧ BAKED ALASKA ❧

Long before the Food Network, Charles Ranhofer emerged as one of America's first celebrity chefs. He did not have his own TV show or a top-end line of kitchenware to license or any of the modern tools that vault chefs to fame today. But from 1862 until 1896, he presided over the kitchens of Delmonico's, the most renowned restaurant in New York City in its day. Ranhofer invented several new dishes, made others, such as Baked Alaska, famous, and changed the face of American cuisine.

THE DARK SIDE OF THE GILDED AGE

The elaborate dessert, Baked Alaska, and Delmonico's restaurant, its equally sumptuous New York City home, evoke America's Gilded Age—or at least one side of it. This era witnessed the creation of enormous personal wealth by captains of industry—such as John D. Rockefeller, Cornelius Vanderbilt, and Andrew Carnegie—and the emergence of the United States as a great industrial power. But there was a dark side to the flamboyant excesses of the age: widespread poverty, injustice, and dramatic social upheaval appeared in stark contrast, like the sooty flip side of a highly polished gold coin.

It is easy to see which side of this coin Delmonico's restaurant catered to, although it was started by two Swiss emigrants, John and Peter Delmonico, in 1827 as a simple pastry shop on William Street in lower Manhattan. Their talented nephew Lorenzo joined them five years later, and by 1837, a redesigned restaurant served gourmet French food in ornate surroundings to New York's growing elite. Delmonico's introduced Americans to the European concept of ordering plates à la carte, or separately from the menu, rather than just being served whatever an innkeeper's cook had on hand that night.

They prepared the elaborate and festive dinner welcoming Prince Albert, the Prince of Wales, on his visit to the United States in 1860. Mark Twain, "Diamond Jim" Brady and his paramour, Lillian Russell, J. P. Morgan, Charles Dickens, and the "Swedish Nightingale," Jenny Lind, were just some of the famous contemporaries who dined there during the Gilded Age.

Chef Charles Ranhofer introduced the latest in French epicurean fashion when he joined as *chef de cuisine* in 1862. Famous dishes such as Lobster Newburg, Chicken a la Keene (which has been renamed and gloppified as chicken a la king), and Baked Alaska originated during his regime.

Born in France in 1836 and formally trained in Paris, Ranhofer served as the private chef for the Duc D'Alsace as a young man. He traveled to America with his next employer, the Russian consul. Lorenzo Delmonico, an already well-established restaurateur, then hired the talented twenty-six-year-old, and the reputations of the chef and the restaurant grew from that time on. Aside from a three-year hiatus in the 1870s, Ranhofer spent his entire career at the legendary Delmonico's.

Ranhofer was a clever marketer as well as an extraordinary cook. He named new dishes (and renamed existing ones) after famous people or events to create some of the first celebrity branding in the culinary world, as well as to drive demand. For example, he christened a potato dish "Sarah Potatoes" after the hugely popular actress Sarah Bernhardt. "Peach Pudding a la Cleveland" recognized President Grover Cleveland, a man of prodigious appetites. Ranhofer even modified an elaborate frozen meringue dessert from France and, with great flair, presented it as a "new" American dish, Baked Alaska.[1]

Baked Alaska exquisitely illustrates Ranhofer's gift for cultivating Americans' excitement and passions through food. The very mention of Alaska conjured up a hotly debated topic at the time. In 1867, facing a crisis in the royal treasury, Czar Alexander II of Russia opened negotiations to sell this "frozen frontier" to the United States. Secretary of State William Seward eventually agreed to the purchase price of $7.2 million, or a few cents for every acre.

Some people—like Horace Greeley, renowned editor of the *New York Tribune*—sneered at the proposed acquisition. Greeley wrote, "Except for the Aleutian Islands and a narrow strip of land extending along the southern coast, the country would not be worth taking as a gift." Others referred to the huge northern landmass as "Seward's Folly" or "Seward's Refrigerator." But after almost a year of rancorous debate in Congress, the United States formalized its purchase of Alaska in 1868, increasing the nation's size by 20 percent with one flourish of a pen.[2]

Along with a frisson of political controversy, the name "Baked Alaska" promised a deliciously intriguing paradox: something cold, possibly frozen, yet somehow counterbalanced by heat. In fact, Baked Alaska is exactly that: an ice-cream-filled cake covered in meringue, which is quickly browned in a hot oven immediately before serving. It represents the aspirational, the luxurious, and the flamboyant—all themes of dining at Delmonico's, the epicurean centerpiece of the Gilded Age.

In the 1860s, creating Baked Alaska was no easy feat, since each of the building blocks of this "bombe" required significant time, elbow grease, and hard-to-achieve temperature control. (A quick aside: Rhode Islander Turner Williams patented the hand-cranked double eggbeater in 1870, a big improvement on the wire whisk but still not what most cooks today would envision as a labor-saving device. And of course ice cream machines were not electrified until the following century.)

In the Gilded Age, Baked Alaska required a highly skilled chef, a group of strong-armed kitchen workers, and the latest in kitchen technology. Today, it is still an impressive dessert, but more of an assembly project.

MODERN BAKED ALASKA

This is more about assembling ingredients and timing it right than showcasing your culinary finesse. My daughter Lucy, now a professional chef, made this dessert when she was nine, although I assisted with the meringue. You can alter the flavors of the cake, the ice cream, and any liqueur according to your taste. I used strawberry ice cream because it is strawberry season as I write this, and I bought lovely fresh strawberries for a garnish. I used traditional butter cake, but you could use any flavor that goes well with the ice cream you have chosen. I think chocolate cake would be delicious with coffee or mint ice cream. I froze the cake in a 2-quart bowl, which makes about 8 servings. You make this bigger or smaller by the size of the bowl you choose. Just make sure the bowl isn't shallow and fits into your oven.

Note on timing: Real work time is about 35 minutes, but the cake needs to be frozen at least 2 hours. It can stay in the freezer, well-wrapped, for up to a week before meringue is applied and browned.

Note on equipment: 2-quart glass or ceramic mixing bowl, heavy-duty baking sheet or broiler pan, aluminum foil, electric mixer.

INGREDIENTS

► 2 moist loaf-shaped pound cakes

► 2 tablespoons rum, brandy, or flavored liqueur (optional)

► 2½ pints highest-quality ice cream

► 4 egg whites, ideally at room temperature

► 1 pinch cream of tartar

► ¾ cup confectioners' sugar plus 3 tablespoons for finishing

DIRECTIONS

Generously line a 2-quart bowl with aluminum foil, allowing enough foil to hang several inches over the edges, so that it can fold over the top of the bowl and cover it once the bowl is filled. Spray the inside of the foil-lined bowl with cooking spray to make the frozen cake easier to get out of the bowl. Remove the ice cream from the freezer so it will be soft enough to scoop once the cake prep is complete.

Slice one loaf cake into half-inch slices. Line the bowl with these, starting with the bottom and overlapping the slices a bit. Cut more slices from the second loaf as needed to fill in gaps and completely line the bowl, with no foil visible. Press the slices into the sides of the bowl with your hands so the inside of the bowl looks like it is made of cake.

Sprinkle or brush the rum or other liqueur onto the sides of the bowl-shaped cake. Use the liqueur sparingly because the cake will not be cooked, meaning the alcohol will be strong.

Scoop the ice cream into the cake-lined bowl and pack in place firmly. Fill the bowl with ice cream, making sure there are no air holes, and smooth the top so that it is flat. Carefully place one layer of half-inch cake slices on top of the ice-cream-filled bowl. Cut wedged pieces to fill in corners, making sure there are no gaps. Press down with fingers or spatula to completely cover ice cream.

Draw up the loose ends of the foil and fold over the bowl. If needed, wrap more foil over the top of the bowl and seal tightly. Freeze for 2 hours minimum or up to one week. Put the broiling pan or cookie sheet in the freezer (if it will fit) the day you plan to serve the dessert.

BEFORE SERVING:

Preheat oven to 450°F. Place the rack in the lower third of the oven to allow enough room for your creation. Beat the egg whites with a mixer at medium speed until white and frothy. Increase the speed to high, and add the pinch of cream of tartar and confectioners' sugar 1 tablespoon at a time. Beat until the egg whites hold stiff peaks.

Remove the ice cream cake from the freezer and unwrap the foil. Lift the cake out of the bowl by pulling up on the sides of the foil. Invert the bowl over the baking sheet, and the cake should come out easily. (Wrap the outer sides of the bowl with a hot, damp dish towel if the cake is stuck. The meringue will hide any imperfections.)

Frost the cake with the meringue, working quickly. Coat the cake with one layer, covering it completely. Add swirls or another layer. The meringue should be very thick. Put the meringue-coated cake in the oven and bake for 3½ minutes or until the meringue is a golden brown. Slide the cake carefully onto a serving platter, using two spatulas if necessary.

Garnish with fresh berries, or if you know your way around a kitchen and enjoy drama with your spirits, present Baked Alaska flambéed with burning rum.

Serve immediately to great applause and enjoy Gilded Age luxury.

Serves 8, if 2-quart bowl is used.

To flambé any brandy, rum, bourbon, or liqueur: Add ½ cup spirits to a small saucepan and heat. Have a pack of long matches or a grill lighter at hand. Have a helper dim the lights around the table while you work on the sauce. Just when the liquor starts to steam, light a match and touch it to the vapors at arm's length. Stand back. *Bring the flaming pan to the table and theatrically pour it over the Baked Alaska or other pudding. Expect marriage proposals from either gender.*

BITE 52

❧ OYSTERS ❧ ROCKEFELLER

New York City was not unique in offering elaborate meals to wealthy clientele during this era of extravagance. Antoine's Restaurant in New Orleans prided itself on new dishes with luxurious flavors. In 1899, chef Jules Alciatore created Oysters Rockefeller, whose name and rich green sauce referenced the enormous wealth of John D. Rockefeller. Along with Andrew Carnegie, J. P. Morgan, and others of the so-called robber barons, Rockefeller was one of the well-known industrialists and financiers who created a dynastic fortune during the Gilded Age.

Jules's father, Antoine, had left France in 1838, hoping to open a restaurant in New York City. A lack of success there prompted him to change his plans, and he settled instead in New Orleans, the largest city in the South and a booming port with a palpable (and audible) French heritage. He opened a popular restaurant and boarding house, which eventually moved to St. Louis Street in the lively French Quarter, where Antoine's still stands today.

Although the French émigré and his wife spent almost forty years successfully running their elegant restaurant, their son, Jules, made Antoine's

In New Orleans, Antoine Alciatore founded his famous Antoine's Restaurant in 1840, soon after Delmonico's opened its elegant doors up north. Fourteen dining rooms, of various sizes and themes, expressed the concept presented by all the era's grand restaurants, where taste in atmosphere and decoration joined taste in food and wine as important parts of the total sensory experience. Unlike Delmonico's, which changed ownership several times after 1919, Antoine's is still family owned and managed by the Alciatore family and continues to serve its celebrated specialty, Oysters Rockefeller.

famous on a national level. Photographs from the Gilded Age show the young, French-trained chef, his family, and his staff outside the restaurant in 1885 and provide glimpses of the ornately decorated dining rooms, which were similar to the draped and gilded Delmonico's of the same era.

Judging from these fine restaurants and their celebrity chefs, it would be easy to presume that the late nineteenth century was one big success story. However, the Gilded Age was not just one lengthy period of growth and opportunity. Economic booms and busts cycled through the nation during this period, and Louisiana was no exception. Racial and ethnic violence, with large-scale lynchings and deadly riots, burned into the country's psyche. (The U.S. Supreme Court upheld Louisiana's "separate but equal" laws in its 1896 decision, *Plessy v. Ferguson*, a ruling that provided Constitutional protection for state-sanctioned apartheid in the United States until 1954.)

Despite financial instability and widespread social injustice surrounding it, Antoine's thrived in New Orleans. Its presence helped establish this city as one of the food capitals of North America. The uniquely delicious ethnic culinary traditions of the local French Creoles and the Acadians (Cajuns)—combined with local seafood and other luscious ingredients—provided a platform for creative chefs who continue to make this fascinating city a destination for food lovers.

Today there are several outstanding restaurants in New Orleans with international reputations. The oldest is Antoine's, which has delighted its customers with more than three million servings of Oyster Rockefeller since its Gilded Age invention by the original Antoine's son, chef Jules Alciatore.

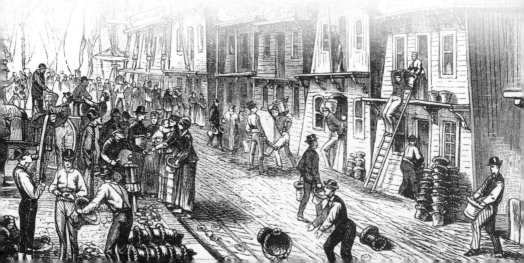

BITE 53

❧ BEEF TENDERLOIN ❧

While Baked Alaska draws oohs and ahhs, and customers rave about Oysters Rockefeller, beef plays a starring role in American cuisine, and has done so for more than three hundred years.

Columbus brought the first cattle to the New World on his second voyage, in 1493, along with hogs, horses, sheep, and other domestic animals. The cattle thrived in New Spain, where huge herds grazed on the open ranges in Texas and northern Mexico. The Mexican herdsmen, known as *vaqueros* or cowboys, developed many of the traditions and much of the vocabulary—rodeo, chaps, lariat, and even ranch—that we associate with the archetypal cowboy of the American West.

Meanwhile, early settlers in the British colonies of Virginia and Massachusetts also shipped cattle across the Atlantic from their European homelands, where the beeves, as they were called, thrived after a few hits and misses. By the time of the American Revolution, there were large herds in every colony, and scientific farmers like George Washington sought the best breeds for local habits. Hogs, sheep, chicken, fish, and game were cheaper and easier to raise. But the early generations of Anglo-Americans

---

Humans have been eating beef for thousands of years. Tribes in the Middle East and India first domesticated cattle around eight thousand years ago, but hunters brought home carcasses of wild aurochs, the big-boned ancestors of today's cows, long before that. Aurochs silhouettes decorate the interiors of ancient cave dwellings in France and Spain, some painted as early as twenty-five thousand years ago. Around the time North America was colonized by the European powers, the English ate more beef than other Europeans, and their consumption was considered a sign of England's prosperity. For most people in Africa, Europe, and Asia, beef was only eaten at a special occasion, like a wedding or a religious festival.

ate a lot of beef, especially in comparison to their Old World continental contemporaries.

Eventually, the Spanish cattle bred with the Anglo-American cattle, and created the Texas longhorn, a breed that resembled the ancient aurochs more than the docile Jersey cows that colonial New Englanders valued for their rich milk. After the Civil War, cowboys herded thousands of longhorns on the lengthy overland routes from Texas north to the railroad terminals in Kansas, Missouri, and Nebraska, where they would then be shipped to the slaughterhouses in Chicago and St. Louis.

Although the huge cattle drives were of near-mythic proportions, they did not last long. After a 1,500-mile journey, the cattle were often tough and stringy before they reached their final destinations. Within a few years, ranchers moved the herds to be nearer the railroads. By the end of the 1880s, the practice of long cattle drives was largely over.

By 1880, more than 4.5 million head of cattle were grazing in the vast grasslands in Kansas, Nebraska, Wyoming, Montana, and the Dakotas. As the American population boomed, so did demand for beef. Meanwhile, the American Plains Indians lost their traditional way of life and their primary food source, the bison, through the settlement of western lands by people of European ancestry and the United States War Department's policy of extermination. It is impossible to separate the growth of the beef industry from the decline of the Plains Indians.

American beef consumption climbed as new distribution networks and technology reduced the price of meat. In 1878, Andrew Chase designed the first refrigerated railcar, which allowed dressed carcasses to be shipped directly to distant butcher shops, rather than shipping live cattle. During this time, Americans also developed their own traditions about butchering, which varied from European practices in vocabulary as well as the precise cut.

But all Western nations shared a special cut of meat that was considered the tenderest and one of the most flavorful. Called the *filet* in France and the fillet in Britain, in America it was (and still is today) the tenderloin. And in the Gilded Age, the tenderloin appeared on the elaborate menus of the nation's most lavish restaurants. Places like the Palmer House in Chicago, Sherry's and Delmonico's in New York, Antoine's in New Orleans, and the Palace in San Francisco featured tenderloin steak, often with rich sauces that would seem to overwhelm rather than emphasize the succulence of the meat.

> *About 25 percent of the cowboys were of African American heritage, many of them former slaves who found new opportunity in the West. Mexican cowboys also drove these herds, and some Chinese workers joined as well, often as cooks. Add the fact that the white cowboys originated from all over Europe—from Ireland to Russia—and it is clear just how multinational these cattle drives were.*

At this point in American history, certain lively neighborhoods in various U.S. cities also became known as the "Tenderloin" because they offered juicy morsels of a different kind of pleasure. The story behind the nickname is that one tough police captain, Alexander Williams, was transferred in 1876 from New York City's gang-infested Gaslight District to the theater district of lower Manhattan, replete with gambling dens, saloons, and "disorderly" houses—brothels.

When asked about his reaction to the move, Williams supposedly replied, "Ah fine, my boy, fine. It's been chuck steak for me; it'll be tenderloin now." By 1878, the term "Tenderloin District" had crossed the continent, becoming the label for the area around Powell and Market Streets in San Francisco, where the name still appears on city maps.[3]

BITE 54

COLD CEREAL

In the nineteenth century, most Americans weren't anywhere near wealthy enough to dine on Oysters Rockefeller or Baked Alaska, but they were notorious for eating large breakfasts. Along with oatmeal porridge or cornmeal mush, people consumed eggs, various breads, and a wide variety of meat and fish—from the traditional breakfast sausage to cod cakes to lamb chops—for their early-morning meal.

Nutritional advisers, both medical professionals and self-educated zealots, pointed to the heaping breakfast tables as the source of nationally recognized health problems, such a dyspepsia and nervous declines. (Every era, it seems, has its defining health complaints about our diets…acid reflux, anyone?) And a group of remarkably colorful characters developed alternatives to the meat-rich American diet by encouraging their countrymen, young, old, and in-between, to eat whole grains and more vegetables. From their messianic point of view, dietary reform had the potential to solve all sorts of problems, from prostitution to political corruption. American's conflicted relationship with food is nothing new.

Sylvester Graham (1794–1851) led this parade early on, inventing "graham bread" in 1829 a few years after his ordination as a Presbyterian minister. His recipe included coarsely ground whole-wheat flour that was free from chemical additives such as alum or chlorine, which commercial bakers used to whiten their bread. This was a radical departure from what Americans were used to. To aspiring families, white flour represented not only purity, but refinement, exactly what they were looking to achieve in the Gilded Age.

> *Americans weren't the first to buy into the idea that white bread was better. In Western civilization since the days of ancient Rome, people from all backgrounds associated soft white bread with upper-class eating habits. The whiter the bread, the better.*

To combat this rising clamor to keep up with the Joneses, dietary reformers like Graham urged the consumption of hearty, unrefined bread, whole meal crackers (hence the invention of graham crackers that are still popular today), and a cold cereal that Graham called "granula"—the origin of our popular granola. Combined with a vegetarian food regime and abstinence from alcohol and other stimulants, Graham's recommended diet would supposedly reduce the excessive sexual desire that irritated Americans' bodies and left them open to a variety of diseases. Masturbation—a prime evil in Graham's religious doctrine and, he argued, a leading cause of blindness—would also be curbed by these health reforms, and so would lustful behavior of all kinds.

While contemporary writers and editors made fun of Graham's zealotry, and some commercial bakers were furious, he had a remarkably widespread influence on the American diet. This was due in part to the lower cost of printing on the newly invented mechanized presses at the time, which made it cheaper and easier for Graham to distribute his tracts touting such an extreme diet. Plus the revolutionary technology of the railroad expanded his ability to address new crowds across the country by traveling on the popular lecture circuit. As historian Cindy Lobel has commented, Graham and other diet reformers "stood on a soapbox of unprecedented height."[4] They were, in essence, the first successful disseminators of fad diets on a national scale.

Operated by the Seventh-Day Adventist Church, the Battle Creek Sanitarium was a highly regarded facility for people looking to detoxify their systems and live healthier lives. Kellogg, a skilled surgeon who regularly treated indigent patients for no fee, attracted a remarkable range of clients during his lifetime, including Amelia Earhart, William Howard Taft, Henry Ford, and George Bernard Shaw.

Diet reformers only grew in numbers and fame after the Civil War. John Harvey Kellogg, MD, (1852–1943) served as the chief medical officer of the Battle Creek Sanitarium in Michigan, a health center emphasizing exercise, a low-meat, mostly vegetable- and whole-grain-based diet, and frequent enemas that were part of the water cure offered there. According to Dr. Kellogg, lustful living was a sure road to a short, unhappy life, and masturbation was particularly harmful. Reforming your diet could change your world. Thus, for him and his patients, alcohol and caffeine were out, yogurt and granola (or "granula" as Dr. Kellogg called it) were in.

Providing fertile ground for the swelling tide of reformers was the strong underpinning of evangelical Christianity that characterized so much of antebellum American popular thought. People sought ways of improving not only their own moral stance but that of society as a whole. Some of these reformers addressed the major issues of their time, like slavery, child exploitation, and women's rights. Others fought against white bread. From Graham's point of view, a vegetarian and whole grain diet improved one's chances of living a saintly life and would make America a better place.

But John Harvey Kellogg wanted to spread his doctrine beyond Battle Creek. In 1897, he and his younger brother, William Keith Kellogg, started a whole-grain cereal company with the goal of improving the American breakfast by simplifying it greatly and making it entirely grain-based, along the lines of granola but even plainer. Arguing over the recipe—William wanted to add extra sugar to the toasted flakes of cereal to make them more palatable—and the use of the Kellogg name, the brothers went their separate ways.

An advertisement for Kellogg's corn flakes.

Ironically, it would be the younger brother's company that would go on to realize the older brother's vision, producing Kellogg's Toasted Corn Flakes and becoming an international cereal behemoth. The extra sugar in William's version won over new customers, who ate their morning cereal in the happy belief that this was not only modern and tasty, but also a righteously healthy way to start the day.

John Harvey's mission was picked up by one of his patients as well. Charles William Post started his own cereal company inspired by the dietary products at Battle Creek Sanitarium. His first product, the popular Postum cereal beverage, provided a noncaffeinated substitute for coffee or tea, and his cold cereal, Post's Grape-Nuts, followed soon after. In 1904, Post created his own brand of corn flakes. He died in 1914, one of the richest men in the United States. John Harvey Kellogg, who lived until 1943, always believed that Post had stolen his recipe.

While other diet reformers in Europe and America created their own style of whole-grain cold cereal, Kellogg and Post (now owned by General Mills) developed the first mass market for this product. Effective advertising campaigns made cold cereal appear easy, safe, and the choice of good mothers. The simultaneous development of the first national grocery-store chain, the Great Atlantic and Pacific Tea Company, now known as A&P, helped place a box of cold cereal in kitchen pantries everywhere. For better or worse, the legacies of Sylvester Graham and John Harvey Kellogg are still staples in the American diet today.

BITE 55
CRACKER JACK

Not only did Americans like their big breakfasts, they loved their snacks. Cracker Jack was one of the first mass-produced snack foods in America, the land that invented mass-produced snack food. People of course ate snacks from time immemorial. Apples, a bit of cheese, nuts, biscuits, jerky—humans enjoy and often need a bit of something between meals.

In the 1800s, doctors and diet reformers railed against the habit of snacking, claiming that it spoiled appetites and led to indigestion, particularly the dyspepsia that apparently haunted every adult male in the Victorian era. John Harvey Kellogg, for example, believed that snacks presented certain harm to those who indulged, despite the fact that during the Gilded Age, many snacks were primarily what today we would consider wholesome, natural foods.

A young German immigrant named Frederick W. Rueckheim started a small popcorn and candy business in Chicago with his younger brother Louis a year after the Great Fire of 1871. F. W. Rueckheim & Brother opened their enterprise with a hand popper and one pot for heating molasses out of a back room. The brothers experimented with various ingredients until they lighted on a mixture of popcorn, peanuts, and molasses.

The ingredients were inexpensive, and with some careful tweaking, the brothers figured out a way to keep the snack from clumping together into an awkward mess. They knew they had a hit on their hands at the Chicago Columbian Exposition of 1893, one of the great world's fairs of the Gilded Age, when visitors bought the confection by the bagful.

In those days, "crackerjack" was a popular slang term for something outstanding and fun. It seemed like the perfect moniker for the popcorn-peanut mixture. So in 1896, the Rueckheims had the name trademarked and started advertising their product in New York and Philadelphia as well as Chicago.

But the brothers faced a major problem in the warm weather, when shipped Cracker Jack became too sticky and soggy by the time it reached consumers. Fortunately, a friend of theirs named Henry Eckstein had the idea, based on an innovation he had seen in Germany, of using individual bags wrapped in doubled-layered waxed paper inside serving-sized, sealed boxes. The new packaging solved the problem, and the candy company was renamed Rueckheim Brothers & Eckstein.

People bought Cracker Jack from street vendors, drugstores, and, starting around 1907, at baseball games. A year later, the now-famous "Take Me Out to the Ball Game" song mentioned Cracker Jack, and the snack food became an integral part of America's pastime.

To entice new customers, the Rueckheims added a premium coupon in every pack in 1910. The Quaker Oats Company had built its enormous market share for hot cereal in part by offering coupons in their signature cylindrical box. Inspired by that success, the Rueckheims copied Quaker's tactic, then changed it to a little toy. Children everywhere were delighted.

The popcorn, peanut, and molasses treat became America's first mass-marketed snack—some say, its first junk food. By 1947, annual sales reached one hundred million packages.[5]

Now more than one hundred years old and owned by snack-food giant PepsiCo, Cracker Jack still offers a toy surprise in every pack and the promise of a good time at the ball game regardless of who wins.

BITE 56

CHICKEN PAPRIKASH

The changing waves of immigration to the United States serve as an important theme of this book, since each immigrant group brought its own food heritage to add to the big stew pot of American cuisine. Toward the end of the 1800s, however, the nature of immigration shifted. Central and eastern Europeans began to stream into the United States, where they were often met with hostility and prejudice.

Even though they arrived with the same dreams as the earlier settlers, they seemed remarkably different in their domestic culture and appearance, with little or no education, limited English, and impoverished origins. From 1870 to the rumblings of World War I in 1913, more than eleven million of these new immigrants crossed the Atlantic.[6]

Some of these eastern Europeans traveled far into the North American continent, drawn by the Homestead Act that promised 160 acres to any man (citizen or citizen-applicant) who would farm unclaimed land for five years. Willa Cather, in her brilliant novel of Nebraskan farm life, *My Antonia*, writes of the exoticism of Antonia's parents' household, burdened and enriched by Slavic traditions, and so different from the young narrator's Anglo-American family. Some of these families built warm sod houses on the northern prairies, the late-nineteenth-century equivalent of starter homes, while they began the backbreaking work of homesteading.

Others, often single men, sought work in heavy industry, hoping to earn enough money to return home and buy a farm back in their old country. Ethnic prejudice, however, weighed heavily against the new immigrants in many industries. "Native" Americans assigned the new arrivals to the worst, most dangerous jobs. At the Homestead Steel Works, the big steel factory owned by Andrew Carnegie outside Pittsburgh, Hungarians and Slavic workers received the lowest wages and, as unskilled and semiskilled laborers, worked the longest, hottest shifts—often twelve hours a day in grueling conditions. Both at work and in company housing, "foreigners" were segregated from "Americans."

Management exploited these ethnic divisions to reduce the potential for labor unions and strikes. The great labor strikes of the nineteenth century occurred during the last decade, and the one at Homestead in 1892 stands out for its protracted violence on both sides.[7]

Married couples ran boarding houses for their own countrymen, with the wife working the longest shift of all, preparing meals, doing the laundry and cleaning, and watching the children. Children were expected to tend the small vegetable garden and the chickens raised in the small yard near the privy—there was no indoor plumbing—and help with the laundry and cleaning. Schooling was limited. Adolescent boys were sent to the mills or the mines, and their wages helped support their family. Girls hired out as domestic workers or made it possible for their parents to take in more boarders by shouldering more cooking and cleaning duties at home.

The boarding house kitchen was a beehive of activity all day. A Hungarian wife, for example, might serve her boarders a filling breakfast of eggs, bread, a little sausage, and coffee. The workmen ate soup for lunch, because it was rehydrating and easiest to eat—an important factor when there was no free time for meals during their twelve-hour shifts. Dinner might be potatoes, spaetzle, or *nokedli* (Hungarian dumplings), root vegetables or beans, and cheap cuts of meat, stewed for hours and served with bread. The classic Hungarian dish, *gulyas* (goulash) incorporates many of these ingredients, and the amount of meat in the pot often reflected the household's income, which was usually minuscule.

Unlike the northern European workers who ate a bland diet, Hungarians enjoyed lots of spices in their food, especially paprika. Even without special seasonings, food was expensive in the factory towns, eating up about 40 percent of an unskilled workman's weekly wage of five dollars. So the amount of meat in the pot might be scant indeed.[8]

Still, there were festive moments in the year, like a daughter's wedding, a christening, or Christmas, when the household cook might prepare something special for her family and her boarders. She and her husband might open some wine for their guests to enjoy with the meal and during the inevitable dancing. A traditional Hungarian dish like Chicken Paprikash—easy, delicious, and not too costly—might grace the dinner table, bringing flavors from home to the hearts of the hardworking new immigrants who helped build America.

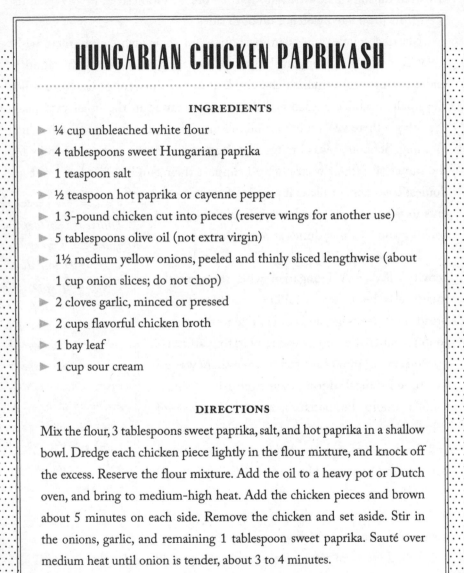

HUNGARIAN CHICKEN PAPRIKASH

INGREDIENTS

- ¼ cup unbleached white flour
- 4 tablespoons sweet Hungarian paprika
- 1 teaspoon salt
- ½ teaspoon hot paprika or cayenne pepper
- 1 3-pound chicken cut into pieces (reserve wings for another use)
- 5 tablespoons olive oil (not extra virgin)
- 1½ medium yellow onions, peeled and thinly sliced lengthwise (about 1 cup onion slices; do not chop)
- 2 cloves garlic, minced or pressed
- 2 cups flavorful chicken broth
- 1 bay leaf
- 1 cup sour cream

DIRECTIONS

Mix the flour, 3 tablespoons sweet paprika, salt, and hot paprika in a shallow bowl. Dredge each chicken piece lightly in the flour mixture, and knock off the excess. Reserve the flour mixture. Add the oil to a heavy pot or Dutch oven, and bring to medium-high heat. Add the chicken pieces and brown about 5 minutes on each side. Remove the chicken and set aside. Stir in the onions, garlic, and remaining 1 tablespoon sweet paprika. Sauté over medium heat until onion is tender, about 3 to 4 minutes.

Lower heat to a simmer. Add the chicken pieces, chicken broth, and bay leaf. Cover tightly and allow to simmer about 30 minutes over very low heat, which will keep the chicken from getting tough. Mix ½ cup sour cream thoroughly with the reserved flour. Add a ladle full of liquid (about ¼ cup) from the pot to the sour cream mixture, and whisk until smooth. Stir the mixture into the pot slowly and simmer for 5 minutes.

Remove from the heat. Stir in the remaining ½ cup sour cream. Keep warm! Serve on noodles or Hungarian-style dumplings, known as *nokedli*.

Serves 4 to 6.

BITE 57

★ SCRAPPLE ★

Despite the rapid development of factories and urban life in the late nineteenth century, most Americans still lived on farms. The growth of the cities, powered by trade and rising industry, created a hungry marketplace for agricultural produce so farmers' goods were in high demand. Technological developments, like the McCormick reaper and the John Deere tractor, made it possible to cultivate the great prairies, once considered untillable, and to reap larger harvests than ever before to feed the growing urban populations.

However, a farm family's prosperity hinged on unpredictable weather conditions, pests, and the health of livestock—and of family members as well. It was always a challenge to save enough to cover the costs of a bad year or two or three. To manage a household economically, women followed traditional recipes that created a second or third meal out of leftovers and table scraps.

Originally from Westphalia, Germany, this pork dish was called *panhas* after the cauldron in which it was boiled, a name it retains to this day among the Amish and some German-speaking sects. The rest of us refer to it as "scrapple," typically made from ground scraps of a slaughtered pig—feet, heart, liver, cleaned intestines, and whatever else is leftover but still edible—and ground grain. Essentially, the pork leftovers are cooked with buckwheat flour, cornmeal, and seasonings. Cooled in loaf pans until it solidifies, the scrapple is then sliced and fried or broiled.

Scrapple provided a tasty, cheap source of protein and carbohydrates in an era when farmers and factory workers burned a lot of calories working out in the fields or in heavy industry. Served with fried eggs for breakfast, in a sandwich for lunch, or with sautéed apples as a supper dish, it was richly satisfying. Pennsylvania, with its strong German agricultural heritage, produces most of the scrapple today, but you can find it in Maryland, Virginia, and the Midwest as well, especially in country diners.[9]

Family scrapple recipes often traveled westward within the memories of emigrating farmwives, who felt no need to write down something so everyday. But thankfully for us, some did. Following are two recipes for Scrapple. One was in print during the Gilded Age, first written by Elizabeth Nicholson, a Philadelphian cookbook author, who published her directions for scrapple in *The Economical Cook and House-Book*. Note that she is still using the word "Indian" for cornmeal. This is the type of haphazard direction that makes you thankful for Fannie Farmer, who standardized measurements and cooking procedures in the 1880s—another Gilded Age achievement.

The second recipe is a modern modification that will *not* make your kitchen feel like a butcher shop. Buckwheat was used on its own with the pork scraps in Westphalia to make *panhas*. German immigrants added the cornmeal soon after their arrival, making it a truly American dish.

TRADITIONAL SCRAPPLE

Take a Pig's haslet (edible organs such as the heart, liver, etc.) and as much offal, lean and fat pork as you wish to make scrapple; boil them well together until they are tender; chop them fine, after taking them out of the liquor; season as sausage, then skim off the fat that has arisen where the meat was boiled; to make all soft, throw away the rest of the water and put this altogether in the pot and thickening it with ½ buckwheat and ½ Indian. Let it boil up, then pour out in pans to cool. Slice and fry it in sausage fat.[10]

MODERN SCRAPPLE

You can find many recipes for scrapple online today, most of which require pig organs and a meat grinder. Here is an easier way for today's hurried cook.

INGREDIENTS

► 1½ pounds bulk pork sausage, preferably without fennel seeds

► 2 cups whole milk

► 1 cup yellow cornmeal

► 1 tablespoon maple syrup or honey (optional)

► 1 teaspoon powdered sage

► ½ teaspoon freshly ground black pepper

► ½ cup maple syrup for serving

DIRECTIONS

Brown the sausage meat over medium-low heat in a large, deep skillet, breaking up the clumps as it cooks. Drain fat. Add the milk to the sausage and bring to a low simmer. Pour in the cornmeal slowly, stirring it into the sausage mixture. Simmer and stir for about 5 minutes. Add the maple syrup or honey, sage, and pepper, stirring to incorporate. Mixture will be very thick. Generously grease one 8- by 4-inch loaf pan. Spoon scrapple into the pan. Wrap tightly in aluminum foil or clear wrap and refrigerate overnight.

The next day, slice the scrapple into ¼- to ½-inch-thick slices. Fry the slices in a lightly greased pan until golden brown on both sides. Serve with eggs for breakfast or simply by itself, with warm maple syrup on the side.

Serves 6 to 8.

BITE 58

Bagels and Bialys

Like scrapple, certain foods vividly resonate in the story of American ethnic groups: spaghetti with meatballs for Italian Americans, potatoes for the Irish, beer and sausage for the Germans, and bagels and lox for the Jews. Like most clichés, there is some truth in these gastronomic associations. But although the first few Jewish immigrants arrived in the 1600s, they came from western Europe, where bagels were uncommon.

It was not until the late 1800s that large numbers of eastern and central European (or Ashkenazi) Jews began to immigrate here. Between 1880 and the beginning of World War I in 1914, almost two million Ashkenazi Jews arrived, in part to escape rampant persecution through the pogroms of their homelands. Along with their meager belongings and rich traditions, they brought their recipes for bagels.

Anti-Semitism was common in the land of the free and the home of the brave, but it never approached the hideous levels prevalent in Poland, Russia, and other countries. These new Ashkenazi immigrants tended to settle in urban enclaves, such as Chicago and New York City. Many Jewish families clustered in the tenements of Manhattan's Lower East Side, home to struggling ethnic groups since the early 1800s. By the mid-1880s, bagel bakeries opened on the Lower East Side to serve the local population, and the dense, round kuchen (small baked goods) developed a following among New Yorkers from all walks of life and religions, which still runs strong today.

Cultural historians debate the origins of the bagel in Europe. One early mention of "beygels" appears in records from Krakow, Poland, in 1610, but the exact history remains in dispute. Experts do agree on the traditional ingredients and process, however. Wheat flour, yeast, and salt were mixed together with water, sometimes with a smidgen of sweetener. After rising, the dough was mixed by hand into small rings, each about four inches in diameter.

The bagels would then be boiled ("kettled") for a short time before baking. The kettling process is what sets bagels apart from other small breads.[11] (Of course, lots of flavoring is added to bagels today: onion, poppy seed, sesame, cinnamon and raisin [my favorite], egg, and so on.) The hole

A man selling bagels.

in the center, which allows for even cooking, also enabled street vendors to display bagels on dowels or heavy cords for their customers.

Until the 1970s, bagels were hard to find outside major metropolitan areas, especially because they went stale quickly. Harry Lender in Connecticut finally set out to fix this problem, developing the first frozen bagel packages. Traditionalists frowned upon this, but Lender's process eventually made bagels available—and widely popular—all over the United States.

Unlike bagels, bialys never achieved national stardom. Named after their home in Bialystok, Poland, Yiddish-speakers called them "bialystoker kuchen." A bialy is a round, soft roll with a depressed center, with minced onions in the middle. Bialys are never kettled, simply baked. They arrived in New York City with the same waves of eastern European Jews who introduced bagels to the United States in the late nineteenth century.

Famed *New York Times* food editor Mimi Sheraton visited Bialystok, Poland, in hopes of finding (and eating) a truly authentic bialy. Sadly, there no longer were any to be had in that city. Its Jewish population had been decimated during the Holocaust of World War II, and after the war the few survivors faced another round of violence. Their traditions survived only through the cultural transplantation of the late nineteenth and early twentieth centuries.

Bialystoker kuchen continue to be hand shaped and baked in Manhattan's Lower East Side at Kossar's Bialys. In true American fashion, the kosher bakery stands on Grand Street near Essex, between a pizzeria and a falafel shop, close to the heart of the historic district that once teemed with eastern European immigrant families. Across the corner, a pickle shop offers old-fashioned sours out of a traditional wooden barrel. Much has changed on the Lower East Side, but if you hurry, you can still enjoy an authentic taste of history.[12]

BITE 59

⋆ CELERY ⋆
(Apium graveolens var. dulce)

Celery boasts a long and venerable history, reaching back to ancient Egypt. Garlands of celery leaves adorned King Tutankhamun's tomb in 1323 BC, and wreaths woven from the leaves crowned the heads of the winners in the Isthmian Games in ancient Greece. The Romans associated celery with fertility and sexual desire.

By the Middle Ages, Europeans used celery primarily for medicinal purposes, as hangover prevention, for example, or to help with insomnia. Later, they used the minced leaves in soups and stews, and grated the bulbous root, celeriac, for a vegetable dish. Its association with fertility meant that celery sometimes made an appearance at wedding feasts. This association lives on at traditional Amish nuptials, where celery plays a decorative and culinary role at the wedding feast.[13]

The celery plant in all its glory (and various stages).

Like their British forebears, Americans chopped celery leaves into soups and stews, or served the diced stalks creamed as a side dish for game. But crunching on a raw celery stick could be a tough, fibrous experience. Then in 1856, George Taylor, a Scots immigrant farming in Kalamazoo, Michigan, began to bank soil around the stems of the celery while it was growing. This blanched the stalks, which made them sweeter and tenderer—and a Gilded Age food trend was born. Kalamazoo became the "Celery Capital of the World."[14]

With the advent of refrigerated railroad cars, celery was shipped all over the country. It was just exotic enough, a somewhat pricey but not outrageous treat, not beyond the means of many Americans but still aspirational. Celery became hugely popular, and the development of special dishes for its presentation illustrates its status during the Gilded Age. Celery-specific tableware appeared on upper-middle-class tables and in fancy restaurants: footed celery vases and long, narrow celery dishes, usually made of cut glass. Although some of these items were handcrafted of crystal, most were produced in factories, a key symbol of industrializing America, which made them affordable for families who were already pursuing the American dream of self-betterment.

Like the presence of a piano in the parlor, celery stalks in their specific service ware indicated that the owner had joined the ranks of the comfortable middle class. The footed vases created for celery and filled with the lush green stems still carrying their leaves, provided an edible decoration for the heavily laden Victorian table. The elongated dishes for the stalks often sported red radishes as punctuation marks at each end. The easy availability of the celery vases and dishes in antique stores and online today testifies to their popularity over more than one hundred years ago, and to the long tradition of Americans enjoying the prestige of certain consumer items for the table.

Today, Europeans enjoy celeriac salad (celeriac is a root vegetable that is part of the celery family), while Americans commonly encounter the celery stalks minced in our tuna fish sandwiches, served with a dip, slathered with peanut butter and raisins (the popular snack called "Ants on a Log"), or stirred in a Bloody Mary. But celery's true glory days shone during the Gilded Age, when it graced the tables of the rich, the famous, and the large, aspiring middle class.

BITE 60
❧ BARBECUE ❧

Few foods are more American than barbecue. Even the word has a New World etymology, most likely derived from a West Indian term, *barbacoa*. *The Oxford Encyclopedia of Food and Drink in America* defines barbecue as "a method of slow-cooking meat over coals," but techniques vary regionally the way any really old recipe will.[15] Here, I am not talking about the great steer barbecues of Texas or the barbecued mutton from Kentucky or even the slowly roasted goat of the Caribbean, although these are all delicious. I am focusing on southern American pork barbecue, a taste of heaven with a drawl.

When Christopher Columbus imported the first hogs to the New World in 1493, he had no idea how well they would thrive. Hogs multiplied in the vast forests of America, particularly in the American South where semiferal pigs provided a cheap (or, illicitly, free) source of meat to laborers and slaves. The meat was lean and stringy, but slow cooking broke down the muscle and gristle, turning an unappetizing, tough beast into a melt-in-your-mouth feast.

Two hundred years after Columbus, English settlers in the southeastern colonies enjoyed pork barbecue at weddings, holiday parties, and other gatherings. Slow roasting a few pigs guaranteed a good turnout for elections as well. George Washington noted his attendance at an Election Day barbecue in Alexandria, where he spent the night after the feasting (and probably the drinking.)[16]

By the 1830s, during the Age of Jackson, political parties incorporated barbecuing into their campaign tactics. The meat was cheap, and the roasting

> *Some places that offered takeout pit barbecue may have had a diverse customer base, but in much of the South, white-owned barbecue restaurants were often "Whites Only." The strict rules of racial segregation applied to these barbecue joints as well as to trains, hospitals, hotels, and schools for almost one hundred years after the Civil War.*

process took more time than dollars, especially since enslaved men often did the hard slow work, basting and turning the skewered pork over the hot coals for hours, until it was tender enough to slip off the bone. Barbecues also featured as the main attraction at church picnics, with congregation members providing a vast selection of side dishes, including Hoppin' John, a mixture of black-eyed peas and white rice with African roots.[17]

During the Gilded Age, free African Americans, sometimes with generations of pig-roasting experience, set up informal outdoor pit barbecues along the roadside, with blue smoke enticing customers to enjoy the flavors of hardwood coals and pork. Successful barbecue cooks might set up in tin-roofed shacks, where the takeout service attracted both blacks and whites at a time when race mixing was rare.

Traditionally served with cornbread, rice dishes, and collard or beet greens simmered with smoked ham bone, the dressed hog might be rubbed with spices before it was roasted over a slow fire or in a pit. In the Memphis area, a sweet, tomato-based sauce accompanied the "pulled" (shredded) meat, and smoky ribs are still the local favorites, while North Carolina is still known for a bite of vinegar in the slightly sweet, spicy flavoring. Eventually, these culinary traditions would travel northward and eastward

A nineteenth-century depiction of African Americans preparing a southern barbecue.

to those industrial cities that offered job opportunities and more freedom to black Americans.

The sweet, spicy taste of Kansas City barbecue sauce is the most widely recognized as "typical" barbecue flavor, but it is based on the recipes of the southeastern cooks who arrived in Kansas after the Civil War. And the luscious smoky barbecue of Chicago is a tasty reminder of the Great Migration when African Americans emigrated from the South to the North in search of new opportunities during the first half of the twentieth century.

Today barbecue continues as a South-rooted national tradition, with an international fan base. The serious debates focus on the pros and cons of different rubs and sauces, the types of coals required, and even what breed of hog produces the best results. The old-fashioned heritage breeds can be purchased online and are worth the extra money if you want to barbecue a Gilded Age (or Gritty Age) hog.

A HOT PROPOSITION: TABASCO!

The oldest commercially made hot sauce is Tabasco sauce, invented in 1868 by Edmund McIlhenny, a Maryland-born former banker who had moved to Louisiana around 1840. Initially McIlhenny used discarded cologne bottles to distribute his sauce to family and friends, and in 1868 when he started to sell to the public, he ordered thousands of new "cologne bottles" from a New Orleans glassworks. It was in these that the sauce was first commercially distributed.

First Edmund's son John and then eventually his younger son, Edward Avery McIlhenny, a self-taught naturalist fresh from an Arctic adventure, assumed control of the company, running it from 1898 to his death in 1949. Like his brother had, Edward focused on expansion and modernization. Paul C. P. McIlhenny, who died in 2013, was the sixth in a line of McIlhenny men to run the business. Tony Simmons, the current company CEO, is a McIlhenny cousin.

chapter 7

HOT DOGS, LIBERTY GARDENS, AND BATHTUB GIN

The Progressive Era, World War I, and Prohibition (1900–1928)

 he early years of the twentieth century witnessed a period of rapid change in society, technology, and food. Progressive reformers, who gave their name to this era, sought to clean up government on the state and national level after the corrupt excesses of the Gilded Age. When local laws failed in governing national food distributors, progressives urged the federal government to step in. For the first time, Washington, DC, set standards in food purity, which had been under the purview of local municipalities for generations.

During World War I, the government organized large-scale voluntary efforts to build our food supply through conservation and household Liberty Gardens. A year after the war ended and after almost a century of struggle for political rights, women finally won the right to vote through the Nineteenth Amendment to the U.S. Constitution. And in 1920, the same year that women could vote in every state, rigid idealists combined with extremists to ban the consumption of alcohol on a national basis through a constitutional amendment known as the "Great Experiment."

BITE 61

HOT DOGS

Hot dogs provide a vivid platform for introducing the Progressive Era. Today, we count on the federal Food and Drug Administration (FDA) to make sure food products are in fact edible. But at the dawn of the twentieth century, only local or state governments investigated food purity, and many of them weren't doing their job. Progressive reformers, a journalist, and a young U.S. president named Theodore Roosevelt changed that, beginning with a new type of sausage called a hot dog.

For more than two hundred years, Americans from all ranks and ethnic groups ate sausages with gusto. Seasoned minced meat—primarily pork, but also beef, chicken, lamb, veal, or a combination—packed into casings (usually sheep intestines) provided an important form of protein that, purchased at the local butcher's, was easy to prepare and delicious to eat at home. Smoked and dried sausages had the additional benefit of having a long shelf life.

Waves of German and Italian immigrants guaranteed that tasty sausage became a ubiquitous part of the nation's menu. The small sausage known as a wiener, reflecting its origin in Vienna, was very popular. The frankfurter style sausage, originating in Frankfurt am Main, Germany, made an early U.S. appearance in 1893 at the World's Columbian Exposition in Chicago.

Anton Ludwig Feuchtwanger sold hot frankfurters from his cart at the exposition, lending customers white gloves to keep from burning their fingers. When the gloves started disappearing as a handy souvenir, Feuchtwanger served the franks in long, narrow buns. Thus, the first hot dog was invented.

The origin of the name "hot dog" is hotly disputed. In 1901, Chicago cartoonist Tad Dorgan drew an image of a dachshund inside a frankfurter bun. Some etymologists argue that "dog" was also an old slang term for sausage, so a heated sausage evolves easily into a hot dog. But whatever the origin, the frankfurters in narrow rolls sold like the proverbial hotcakes, especially at games and fairs. No knife, fork, or white gloves were necessary. Convenient, neat, cheap, and tasty!

Sales of hot dogs and all processed meats tumbled in 1906 with the publication of Upton Sinclair's *The Jungle*. Meant to expose the horrific working conditions in the Chicago slaughterhouses, the book instead incited readers to protest the filthy environment in which commercial meat-packers processed their product. "I aimed at the public's heart and by accident hit it in the stomach," Sinclair observed.[1]

Although reformers had pushed for more government oversight of food production for many years, the journalist's exposé galvanized the protests. The graphic descriptions of revoltingly unsanitary practices for the creation of such popular foods as bologna and hot dogs broadened and amplified the public outcry for reform.[2] Within six months of the publication of *The Jungle*, President Theodore Roosevelt and other progressives responded with the passage of the Pure Food and Drug Act, which established health standards and, more importantly, required their enforcement by the federal government.[3]

The passage of the Pure Food and Drug Act, with its extension of federal regulatory power, reassured consumers remarkably quickly, and the sales of processed meats, including hot dogs, revived. In 1916, Nathan's built one

A man and a woman enjoying hot dogs at a fair.

of the first hot-dog stands on Coney Island, New York, selling franks on narrow buns. Oscar Mayer, Swift, and Armour each began to market their own brands of hot dogs nationally. As the First World War raged in Europe, Americans looked nervously across the Atlantic, but still enjoyed their red hots piled high with relish, sauerkraut, and mustard, or just plain.

FAIRS ORIGINATE SOME INTERESTING FOODS

Beginning with Prince Albert's Great Exhibition at the Crystal Palace in London in 1851, massive expositions and world fairs were enormously popular with most Americans—from inventors and honeymooners to heads of state and factory workers—for more than one hundred years. These ambitious events presented technological advancements, foreign curiosities, entertainments, and new foods to the delight and wonderment of the visitors, who flocked from all over the country to experience the novel sights, sounds, and tastes.

The Philadelphia Centennial Exposition in 1876, for example, covered 236 acres in Fairmont Park, where more than ten million visitors paid fifty cents a head to visit the thirty thousand exhibits. Exotic foods, such as pineapples and bananas, impressed the attendees and built consumer demand. The 1893 Chicago World's Columbian Exposition launched the hot dog and inspired a young caramel maker named Milton Hershey to create the Hershey Bar using new chocolate-making technology he witnessed there. An impressive number of America's favorite foods first reached consumers at the St. Louis World's Fair in 1904, including Cracker Jack, Heinz Ketchup, and the ice cream cone.

Like many public events in the United States, these fairs were aimed at a white audience and granted African Americans attendance only on certain days, if at all. The expositions demonstrated wondrous technology, agricultural advancements, and delightful new foods, but they also showed the deep racial prejudice that characterized America, North and South, in the Gilded Age and Progressive Era.[4]

Many years later, in 1939, President Franklin D. Roosevelt served Nathan's Famous Hot Dogs from Coney Island to the British king, George VI, and his wife, Queen Elizabeth, at a picnic at the Roosevelts' family home in Hyde Park. The informal meal made international headlines and helped cement Anglo-American relations prior to the outbreak of World War II.[5]

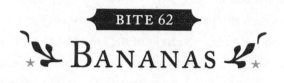

BITE 62
BANANAS

While packaged foods made life a little bit easier for wives and mothers in the Progressive Era, bananas rose in popularity as a tropical convenience food.

We are so accustomed to seeing bananas everywhere in our markets, cafeterias, and fruit bowls that it's hard to imagine life without them. How did they get so popular?

Honduras was only one particularly egregious example of the term "banana republic," a phrase coined by writer O. Henry in 1904.

Originating in Southeast Asia, bananas were cultivated in Africa by 600 AD. From there, they followed the spread of Islam via Arab traders to the Iberian Peninsula and the Middle East. The Portuguese introduced bananas to their New World colonies by the 1600s, and the tropical fruit flourished in South and Central America. Northern Americans infrequently encountered bananas, an exotic fruit that spoiled quickly, until the development of refrigerated storage tanks on rapid steamships in the 1880s. By 1900, the Boston-based United Fruit Company not only controlled vast plantations in tropical Central America but also a fleet of well-equipped steamships and warehouses in America's major ports.

People from all walks of life grew to love bananas. Scientists joined health faddists by touting the nourishing vitamins and the sanitary, germ-free wrapper provided by Mother Nature. At a time when muckraking journalists exposed the toxic additives in many foods, as well as the filthy conditions in some food factories, bananas appeared to be an ideal food for all ages. By 1902, popular cookbook author Sarah Tyson Rorer included several banana-based recipes in *Mrs. Rorer's New Cook Book*.[6]

At the same time that bananas grew in popularity, President Theodore Roosevelt expanded America's military presence in Central America and the West Indies. Roosevelt sent troops in support of American business interests to Honduras, where the United Fruit Company and other U.S. corporations dominated the nation's infrastructure as well as its economy.

That same year, the president declared the Roosevelt Corollary to the Monroe Doctrine, asserting the right of the United States to intervene militarily in the economic affairs of Central America and Caribbean countries. The booming market for bananas, and the closely held supply chain, created record profits for American investors in Central America.

A few years earlier, the recently organized U.S. Immigration Service opened Ellis Island, the Progressive Era immigration depot in New York Harbor. The United Fruit Company stationed an agent at the South Ferry Terminal where the Ellis Island ferry docked in Manhattan. Handing a banana to each immigrant who stepped off the boat, the agent said, "Welcome to America!" although the banana was as foreign to these shores as the men and women who were arriving from the Old World.[7]

A dining hall at Ellis Island, where immigrants had to pass before they could enter America.

First built in the 1890s, Ellis Island witnessed the boom in immigration—50 percent from eastern Europe—prior to the outbreak of World War I. The number of people processed hit its peak in 1907, when over one million new arrivals passed through Ellis Island's gates on their way to America. The average stay on Ellis Island was three to seven hours. Although officials detained many immigrants for medical reasons, only 2 percent were ultimately refused entry and deported. The immigration service provided meals in the vast dining hall, which could serve one thousand people at a time.[8]

Many immigrants were puzzled by bananas, having never encountered them in their homeland, and there was confusion about how to eat the fruit. Elderly Americans, whose stories have been recorded by Janet Levine and other historians for the National Park Service (which magnificently preserves and interprets Ellis Island today as a national shrine), frequently mention their surprise at being served whole bananas during their short stay on Ellis. Vera Guaditsa arrived from Czechoslovakia as a young woman in 1928 and was delayed due to ill health. She recalled being served a banana in the dining hall:

"I never saw a banana in my life…and they served banana. I was just looking at it. And they ate banana with the skin on! Some were Hungarians, some of them were Polish—and I understood what they were saying because the language was similar. [They] never had a banana before. So when the Czech lady came a couple of hours later, I asked her what it was—the banana—and she showed me how to eat it."[9] Today, it seems incredible that people didn't know to peel a banana, since we are introduced to bananas practically as infants these days.

Once an exotic import, then a symbol of corporate imperialism and an immigrant's puzzle, today they are an everyday addition to the household fruit bowl.

BITE 63

COMMERCIAL CANNING

Technological advances in the nineteenth century brought advances to canned foods that would make them a healthy part of Americans' diets by the Progressive Era. In the early 1800s, British and French inventors had developed prototypes for preserving all sorts of food in tin-lined iron canisters (cans) or in sturdy glass jars. Although expensive and sometimes dangerously contaminated with bacteria causing illnesses like botulism, canned foods were especially appreciated by explorers, naval crews, and other seafarers who had few good choices for meat supplies on board their vessels.

THE IRON CHINK

In Washington State, Chinese laborers efficiently cleaned and filleted the salmon to be canned until racial prejudice and shortsighted nativism led to the federal Chinese Exclusion Act of 1882, which resulted in the deportation of Chinese and other Asian immigrants. The technological response to the stunning drop in the labor supply illustrates the old adage that necessity is the mother of invention. The canning factories were having a hard time finding enough workers to replace the Chinese laborers who had been forced to leave America.

In 1903, Edmund A. Smith, a twenty-five-year-old inventor, created a machine that automated the gutting and cleaning process for canning salmon. The machine, which he christened the "Iron Chink," could clean the pink fish fifty-five times faster than experienced human workers. Smith's unself-conscious racism, reflected in the name of the device, was typical of the early 1900s.[10]

It wasn't until years after the Civil War that the general population consumed canned food regularly. The great meat processors in Chicago were among the first mass producers of canned meat and meat products, while the West Coast canneries developed large national markets for canned salmon and tuna fish.

A few years earlier, scientists at Massachusetts Institute of Technology had charted the safest pasteurization times and the correct temperatures needed for canning many foods. Americans respected the scientific method of food preparation and adopted canned (or jarred) food at a fast clip. Flavorful fresh food in season remained desirable, but canned food—especially fruits and vegetables—provided a more varied, nutritionally rich diet at an affordable price. Housewives learned to trust certain brands of canned food, such as Heinz, Campbell's, and Libby's, during the Progressive Era, when health claims and scientific processes were united in colorful national advertising campaigns.[11]

Although Heinz was founded in the late nineteenth century, H. J. Heinz first incorporated his canned food company in 1905. By then, his tomato ketchup was already a hit. (Today, Heinz Ketchup is world renowned and has a healthy 50 percent market share in the United States.) A progressive businessman, Heinz advocated for safe and healthy food processes, and lobbied for the passage of the Pure Food and Drug Act in 1906.

An advertisement for Heinz Pickles. Note they already have their 57 varieties, the same number that is still displayed on their ketchup bottles today.

The brothers Arthur and Charles Libby partnered with Archibald McNeill in Chicago to form Libby, McNeill, and Libby Canned Foods. In the early 1900s, Libby's began to specialize in canned fruits and vegetables, which have graced the tables of middle-class Americans for more than one hundred years. Today, their canned pumpkin and the recipe on the back label enable even the inexperienced cook to produce a reliable pumpkin pie for Thanksgiving dessert.

Campbell's began as a small New Jersey–based canning company specializing in fancy vegetables. In 1897, they developed condensed soup, which proved so successful that the company's name eventually changed to the Campbell Soup Company. Campbell's condensed soup was an innovative food, a time-saving convenience for the Progressive Era woman. Like corporate management at Heinz and Libby's, Campbell's believed that advertising held the key to growth.

Companies began to pour money into marketing support for their brands, thanks in part to a new federal law that allowed individuals and corporations to trademark their unique logos and images. By 1900, American advertisers were spending $95 million a year on advertising, ten times what they spent right after the Civil War. And by 1919, the advertising world poured more than half a billion dollars into product launches and support.[12] For example, Heinz trademarked and promoted its 57 Varieties, a number chosen because H. J. Heinz liked the sound of it. (According to H. J. Heinz Company, their founder thought the number sounded lucky, even though they produced more than sixty products by the time the 57 Varieties campaign was born.)

The Campbell Kids, with happily plump faces, joined the Campbell campaigns in 1904. Grace Wiederseim Drayton, an illustrator and writer, sketched images of happy children onto her husband's advertising layout for a Campbell's condensed soup. The company adored the drawings' appeal and trademarked Mrs. Drayton's contribution. Recognized as the doting "mother" of the Campbell Kids, Grace Wiederseim Drayton drew for the company for twenty years.[13]

Advertising at the dawn of the twentieth century became a powerful influencer in people's food choices. The high quality of the commercial artwork, which today can fetch high prices at auction, appears quite lovely and innocent to modern eyes. Canned foods, bottled beverages, indeed many of the national brands that we still see on store shelves today, were developing their marketing expertise more than one hundred years ago.

BITE 64

⚘PEANUT BUTTER⚘

Like canned food, peanut butter was a convenience that required the right technology to be produced on a large scale. Even before the invention of powered grinders that could turn the ground nuts into a smooth paste however, nutritional experts were interested in peanut butter as a health food. So this popular sandwich spread had a few founding fathers before it was widely adopted by American mothers ready for a cheap, nourishing sandwich spread for their families.

In 1884, Canadian Marcellus Gilmore Edson patented peanut paste, and Dr. John Harvey Kellogg received a patent for a process to prepare nut butters eleven years later. Health enthusiast Kellogg (see Bite 54) sought an easily digestible vegetable protein for his patients and extolled the virtues of peanut butter.

Later, agricultural reformer and scientist George Washington Carver advocated peanut butter as a nourishing food. A former slave, Carver hoped to lift the rural South out of extreme poverty by developing new agricultural products and processes. The destructive boll weevil was devastating the cotton crop in the South and hurting the household economies of many farms. The harsh Jim Crow laws upheld segregation that oppressed the black citizens, particularly in the Cotton Belt, making them overly reliant on cotton and thus vulnerable. Carver's work encouraged small farmers, black and white, to seek self-sufficiency through crops such as sweet potatoes and peanuts.

These were lean times in the national economy. Peanut butter grew increasingly popular among people from all regions concerned about their health and their pocketbook. Mothers all over the country embraced peanut butter as a healthy convenience food that helped them save pennies. It was the Great Depression, and a peanut butter sandwich became a top choice for a money-saving lunch. But Americans' love for the substance was not limited to hard times. After World War II, the entire generation of baby boomers grew up consuming peanut butter and jelly sandwiches with big glasses of milk.

Early in the 1900s, entrepreneurs developed national brands of peanut butter, with Peter Pan peanut butter appearing on supermarket shelves in the 1920s, followed by Skippy in 1932.

The recipe for today's commercial peanut butter belies its health food antecedents. Extra salt and sweeteners join emulsifiers and hydrogenated oils to keep the butter from separating in the jar. To avoid these potentially harmful (and frankly unnecessary) additives, try the peanut butter you grind yourself at stores like Whole Foods. You can always add a little salt when you get home—I do.

◀ BITE 65 ▶

✹ HOME CANNING AND ✹
FOOD CONSERVATION

Food and the Home Front during the Great War

Life was changing rapidly in the Progressive Era, but the pace of change itself accelerated with the outbreak of World War I, a global conflict with an enormous impact on the old world order. It began with the assassination of the Austrian archduke Francis Ferdinand in 1914, when Germany declared war on Russia and France. Fulfilling its treaty promises, Britain then declared war on Germany. For the next two and a half years, as the Great War devastated northwestern Europe, President Woodrow Wilson promised America he would "keep us out of war." But even before we joined the Allies in April 1917, United States citizens responded to the war with an outpouring of food relief and a remarkably effective volunteer food drive.

Wilson appointed Herbert Hoover as the director of the U.S. Food Administration (USFA). Hoover, an energetic engineer, proceeded to organize a massive campaign directed at every household, encouraging families and individuals to conserve food by reducing consumption and by growing their own. Though America stayed neutral before joining the war, this system would provide more food for the war front, where desperately hungry civilians needed supplies. Ultimately, it meant that the United States could send high-quality food to the Allies and our own troops.[14]

How did the USFA convince people to voluntarily cut back on food consumption and start canning homegrown produce? Historian Rae Katherine Eighmey notes in *Food Will Win the War* that food conservation in World War I was the first large-scale social-networking enterprise of the twentieth century. Long before Facebook and Twitter, and even predating radio and television, the USFA connected with social groups, schools, and

A woman examining foods she has canned to conserve during WWI rationing.

farmers, plus food distributors, wholesalers, and retailers, with their message: cut food consumption and increase food production at home.[15]

The federal agency hired talented artists and graphic designers to produce stunning posters, brochures, billboards, and postcards. In an era when even remote counties had their own newspapers, and rural magazines with names like the *Farmer's Wife* reached seventy-five thousand subscribers, the local and the national commentators urged their readers to grow more food crops.[16]

With colorful posters and flyers shouting, "Food is Ammunition—Don't Waste It!" "Can Vegetables, Fruit, and the Kaiser Too!" and "Boys! Girls! Raise Pigs to Help Win the War," the federal government and local agencies urged all ages to reduce waste and help with food production. By 1918, voluntary rationing of meat (beef, pork, or mutton) and white flour lead to Meatless Mondays and Wheatless Wednesdays, so that household meal planning became a patriotic act.

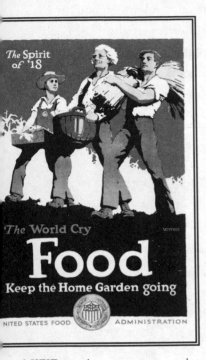

The Spirit of '18

The World Cry

Food

Keep the Home Garden going

NITED STATES FOOD ADMINISTRATION

A WWI wartime poster encouraging people to grow their gardens to support the war efforts.

Meanwhile, federal and state governments urged generous helpings of vegetables and dairy products. The use of meat stretchers, such as bread crumbs and minced crackers, on those days that meat could be served was no longer the province of the poor. Households from all economic strata demonstrated their support for the war effort by eschewing large servings of beef, which no longer made its daily appearance at the dining table.

Historically, one aspect of Americans' diets that differed from the Old World was the quality and quantity of meat. From the plentiful game in the early colonial period to the availability of fresh beef and pork in the nineteenth century, the United States consumed an amount of animal protein that surprised people from other nations. Immigrants from Europe adapted their traditional recipes to include more meat. For example, the ever-popular Italian plate of spaghetti and meatballs was actually developed by Italian Americans, who had significantly more access to beef in the New World, even if they suffered other deprivations. Prior to the USFA's campaign, diners might expect beef and pork at two, if not three, meals a day.[17]

Asking the population to join in on Meatless Mondays without flinching was a tall order from the U.S. government. It went against Americans' daily habits and experience of seeing meat available everywhere and in everything.

But the food conservation campaigns met with a positive response, fed largely by civilians' desire to help with the cause while their sons, brothers, husbands, and boyfriends fought for democracy.

Showing their own patriotic fervor, national publications like *Good Housekeeping* and local journals brimmed with new recipes for meatless meals. Cookbook author Amelia Doddridge published *Liberty Recipes* in 1918, featuring a special section on meat extenders and meat substitutes. These recipes during America's eighteen months of participation in World War I help us understand just how seriously people followed the rationing laws, and how cautiously they avoided wasting food of any kind.[18]

American cooks earnestly developed new recipes for leftovers. The *Farmer's Wife* recommended adding beef gravy to bread crumbs in individual ramekins and adding one egg to each little pot before baking. One food writer suggested adding peanut butter to macaroni for a very inexpensive, high-protein dish.

The USFA also encouraged everyone—rural, urban, or in-between—to grow fruits and vegetables in war gardens, also known as Liberty Gardens, for their own use. Agricultural colleges and farm schools sent out local experts who gave lectures and provided hands-on learning to teach vegetable gardening and food preservation techniques. Locals dug community gardens in urban vacant lots, producing fresh vegetables for their tables and potentially improving nutrition for their families.

The glass canning jars known as Mason or Ball jars came in very handy. John Mason had invented the first Mason jar in 1858, but it took World War I and vastly improved kitchen technology for home canning to hit an all-time high.

Women and older children learned to preserve their excess garden crop, much easier now that home canning equipment was widely available.

The USFA campaign was effective. With widespread household conservation, agricultural production rising, and the increase in food exports, America was able to provide much-needed food for the Allies and for its own troops. Herbert Hoover stopped the rationing efforts a few months before the armistice on November 11, 1918 ended the war to end all wars.

Many people in the national government, including Hoover, now believed that local and voluntary efforts were the answer in time of crisis, since their goals were accomplished without federal enforcement. But bigger challenges awaited that would require a more active government response. And lessons for American families coping with food deprivation would soon be put to use again with the arrival of the Great Depression and the next World War.

THE KITCHEN SOLDIERS

In early 1918, when American soldiers were suffering devastating losses in France and Flanders, widely read *Good Housekeeping* magazine offered a pledge for its readers, "A Gun for the Kitchen Soldier":

> I, the member of the household entrusted with the handling of food, do hereby enlist as a kitchen soldier for home service and do pledge myself to waste no food and to use wisely all food purchased for this household, knowing that by doing so I can help conserve the foods that must be shipped to our soldiers and our Allies.

After signing the pledge and adding their address, women were urged to send the signed pledge to the Good Housekeeping Institute on 39th Street in New York City. The magazine promised in turn to send a certificate "rich with colors, together with the Kitchen Soldier's code."[19]

It should be noted that *Good Housekeeping* worked closely with USFA and proudly kept its readers up to date on the latest ways to support Uncle Sam's efforts. Although we don't know how many pledge cards were actually sent in, we do know that subscription levels remained high during the war years.[20]

BITE 66

❧ DOUGHBOY RATIONS ❧ DURING WORLD WAR I

World War I was the great watershed of this era, the war to end all wars that began with the assassination of Austria's Archduke Ferdinand and Grand Duchess Sophie in Serbia in 1914. By the time U.S. infantrymen (doughboys) arrived at the western front in the summer of 1917, America had been supplying our British, French, and Belgian allies with food *for two years*. The U.S. troops were well provisioned, startlingly so compared to their ally comrades.

That old military standby, hardtack (see Bite 42) was replaced by fresh bread whenever possible, thanks to the development of field bakeries that could provide hot food at the front. Small, wagon-sized food carts, sent from field kitchens located back from the front, even brought hot food into the trenches, where most of the fighting and casualties took place.

Infested with rats and other vermin, subject to poison gas attacks and shelling, and filled with cold, greasy mud, the trenches presented a uniquely harrowing experience for the doughboys, many of them fresh off the farm in America. The arrival of cooked food, along with candy, dairy products, and soft baked bread, struck a much appreciated but incongruous note in the hell known as trench warfare. Of course, there remained the challenge of keeping the food dry, clean, and away from the rats.[21]

Naturally, one big problem for feeding the troops on the front was the safety of the supply lines, which were targets for bombs and other sabotage. Every soldier in the trenches carried emergency rations of twelve ounces of canned meat or fresh bacon, ground coffee, sugar, and tobacco with rolling papers (and later, pre-rolled cigarettes). The army purchased canned meat from the French, which was labeled "Madagascar" and promptly nicknamed "monkey meat" by the Americans in disgust.

In this environment, hardtack still made its appearance on the dough-boy menu. These reserve rations were designed to sustain the troops when the supply lines broke down, or when they were too far from the supply depots. For a generation of men, the servicemen's rations defined part of the wartime experience and the memory remained with them long after they shipped home.

In France, the soldiers were billeted in relative safety before and after their service in the trenches. Here they had dependable access to food and might even receive packages from home. Of course, food shipped from the United States had to remain edible without any extra care. Even when stale and crumbled, any food sent by loved ones was always particularly appreciated.

The American Red Cross, founded by Clara Barton in 1881, played a significant role in this multinational conflict. It not only provided medical care overseas, before there was an enlisted military nursing staff, but also helped on the home front by organizing volunteers and fighting the deadly Spanish influenza plague of 1918 that, worldwide, would kill more people than the war.[22]

The Red Cross also communicated with families about helpful ways to support the troops. Here is a recipe they recommended for folks who wanted to send their soldier a shippable treat.

RED CROSS WAR CAKE

The American Red Cross promoted this cake recipe, promising that the end product could reach the western front and retain its freshness. The dried fruit helps keep it moist if it has to be shipped across the Atlantic. Try soaking the raisins in rum for a few days or a week before you make the cake. Your dough-boy will thank you.

The original recipe comes with a recommendation: "Cake keeps fresh for a long time and can be sent to men at the front."[23]

INGREDIENTS

▶ 2 cups brown sugar
▶ 2 cups hot water
▶ 8 ounces raisins (about 1 package), chopped
▶ 2 tablespoons lard
▶ 1 teaspoon ground cinnamon
▶ 1 teaspoon cloves
▶ 1 teaspoon salt
▶ 3 cups all-purpose flour
▶ 1 teaspoon baking soda

DIRECTIONS

Put sugar, hot water, raisins, lard, cinnamon, cloves, and salt in a large pot. Bring to a boil over medium heat, stirring frequently, then reduce the heat to medium low and cook at a low boil for 5 minutes. Remove from heat and cool. Preheat oven to 350°F. Stir in flour and baking soda. Mix well. Grease 2 mini-loaf pans. Pour batter into the pans and bake for 45 minutes.

Serves 4 to 6.

THEY AIN'T THE PILLSBURY DOUGHBOY...SO WHO ARE THEY?

Why were U.S. Army soldiers called doughboys during World War I? Actually, enlisted men in the army and the marines were called doughboys as early as the Mexican American War of 1846–1848, but it was a somewhat pejorative term at that time. By World War I, the label was used almost affectionately. Its origins are unclear.

The term may refer to the dusty faces of the troops as they marched in Mexico; the dough the troops made for their rations; the pale white faces of newly enlisted men who often hadn't been tanned or sunburned by long treks in the sun; or even the Mexican word *adobo* (a pickle or marinade). In the next generation of American soldiers, beginning in World War II, the doughboy became the GI, which stood for "government issue." GI was also an abbreviation for military equipment made of galvanized iron.

BITE 67

❧ LACE COOKIES ❧ AND OREOS

Women's Suffrage

The role of women in America changed gradually throughout history, but the first twenty years of the twentieth century witnessed important strides in the workplace and at the voting booth. Women played an active role in World War I both on the home front, where they filled the positions left empty by enlisted men and organized food drives, and overseas, where they were nurses, ambulance drivers, and relief workers. More and more, women faced the modern dilemma of working two shifts, at home and at outside work. They were happy to adopt labor-saving devices and foods to make their complicated lives a bit easier.

The Progressive Era saw the rise of premade foods that American consumers found very appealing. As we explored in Bite 63, commercially canned foods delighted a large swath of the public, particularly women, who were happy to skip a step or two in the kitchen with the help of reliably safe tinned stock or canned peas. New advances in household appliances, from eggbeaters and refrigerators to clothes washers and carpet sweepers, made housework more efficient. Stoves and ovens became easier to regulate, even if most of them still burned wood.

Despite all the modern advances, housework continued to require enormous drudgery for those who did not have family or hired help (which were most families at this point). Although some traditionalists at the time might have looked askance at timesavers like commercially made food and new appliances, most women were eager for even a bit of free time.[24]

Many women were also eager for their political rights. The Progressive Era witnessed the culmination of the women's suffrage movement in America. Although the call for women's equality began in earnest alongside the abolition movement in the

Actress Hedwig Reicher performing as "Columbia" during the suffrage pageant in Washington, DC, on March 3, 1913.

early 1800s, the federal government still denied women's right to vote. Led by nineteenth-century trailblazers such as Susan B. Anthony and Elizabeth Cady Stanton, women wrote, organized, protested, and argued for the vote literally for decades.

Like any long-running campaign, the women's suffrage movement was in constant need of funds. The novel concept of the fund-raising cookbook presented one solution. In 1886, Hattie Burr edited a collection of recipes sent in by women devoted to the cause, entitled *The Woman Suffrage*

Cookbook. Sold at suffrage fairs and other gatherings, it raised money for Massachusetts suffragists' campaign for the right to vote in state elections.

Other state suffrage organizations followed suit. Washington had its own cookbook, as did Michigan. The recipes in all these collections ranged from the simple to the complicated and from the traditional to the scientific. One list of ingredients might call for "butter the size of a walnut," while another would specify "two level tablespoons of butter." It was a time of transition in the kitchen, and the recipes reflect that change.

The suffrage cookbooks present an interesting window on their era. As food studies author Emily Contois points out, Hattie Burr's cookbook "communicated with women of all classes in the common language of the cookbook about not only food and domesticity, but also the radical cause of women's right to vote."[25]

On March 3, 1913, more than five thousand suffragists paraded in Washington, DC. Suffrage parades on the state and city level were not unusual, but this was the first time women demonstrated their commitment to the right to the vote on the national stage. The Women's March on Washington stunned and shocked many male (and some very vocal female) contemporaries in 1913, who shouted obscenities and harassed the peaceful demonstrators as they marched down Pennsylvania Avenue. The local police did little to stem the violence, and more than one hundred women were hospitalized.

Although some of the marchers were discouraged, others felt that their treatment helped galvanize new supporters. American suffragists looked to their sisters in Great Britain and found inspiration for steadfastness. Letter-writing efforts continued, but increasingly the leaders sought media support and organized grassroots outreach. They organized a drive across the country, rallying followers to join their campaign.

Women in western states already had the vote, but the rest of the country lagged behind, particularly in the South. A young feminist named Alice Paul led their efforts. (Later, Paul would found the National Women's Party.) By 1916, activist women had chained themselves to the White House fence, demanding the vote. They were arrested and jailed. Many of these prisoners participated in a hunger strike and one died. The government responded by force-feeding the women, a brutal and painful process that created a public relations disaster for the White House.[26]

A year later, in 1917, the United States joined the Allies in the Great War. Women continued to march for the vote, some carrying posters that compared President Woodrow Wilson to the German kaiser for denying democratic participation. Their opponents called them traitors, but the photographs of the posters ran in newspapers all over the country, heightening awareness of the suffrage cause.

Meanwhile, women activists also supported the war effort as factory workers, medical workers, and volunteers. At least 20 percent of workers in wartime electrical-machinery, airplane, and food industries were women. Their dedication helped sway formerly resistant legislators, convincing them that women could actually handle civic responsibilities.[27]

Finally, in 1919, Congress passed the Nineteenth Amendment, guaranteeing women's right to vote. The amendment was ratified in 1920, allowing women all over America to vote in the presidential election between Warren G. Harding, James M. Cox, and Eugene V. Debs. The suffragettes' critics, who declared that voting women would neglect their families and abandon the kitchen, were proved wrong. But women who worked inside and outside the home continued to shoulder the burdens of housekeeping and cooking.

CHOCOLATE FLAVORED WAFERS WITH CREAMY LEMON OR VANILLA FLAVORED FILLING

The development of prepared foods provided at least some time and freedom from the hot stove. During this era, the National Biscuit Company (today known as Nabisco) started selling a range of packaged cookies, including Oreos, which were first produced in their Manhattan factory in 1912. We can only imagine how some mothers saw these chocolate sandwich creations: not as mass-produced, sugar-laden menaces, but as exciting, safe treats their children would love. And mothers themselves would love the little extra time saved simply by opening a package.

OATMEAL LACE COOKIES

This recipe is similar to directions found in the suffrage cookbooks sold to raise money for the cause.

INGREDIENTS

- ► 1 cup light brown sugar
- ► ½ cup butter (1 stick)
- ► 1 egg, lightly beaten
- ► ½ teaspoon vanilla
- ► 2 tablespoons all-purpose flour
- ► ¼ teaspoon baking powder
- ► ¼ teaspoon salt
- ► 1 cup oatmeal (not instant)

DIRECTIONS

Preheat oven to 350°F. Mix the sugar and butter; cream well. Add the egg and vanilla. Mix the flour, baking powder, salt, and oatmeal in a separate bowl, and stir into the butter mixture, making sure all ingredients are well combined. Cover a cookie sheet with parchment paper or aluminum foil. Drop ½ tablespoon of batter per cookie onto the sheet. Leave plenty of room between cookies because they spread a lot. Bake 8 minutes; edges should be brown. Slice cookies apart, if necessary, while they are still warm on the cookie sheet. Cool well before removing cookies from foil. They should be thin, crispy, and delicious.

Store in tins, which you can bring to the election booths and share with your fellow voters.

BITE 68

❧ Cocktails and ❧ the Roaring Twenties

Cocktails sport a long history in America. Although the word first appeared in print in the *Morning Post and Gazeteer* in London, England, in 1798, it didn't really catch on until it appeared five years later in a small pamphlet in the United States. Cocktails called America home from then on. But as we have seen in earlier chapters, people have been mixing wine, hard cider, or distilled liquors with various ingredients literally for ages. Punches, toddies, and syllabubs had strengthened hearts and fuddled heads since the days of the early colonists. The nuances of what makes one mixed alcoholic drink a cocktail and another something by a different name is a bit elusive except for the presence of bitters in the mix.

In 1806, an inquiring mind asked the *Balance and Columbian Repository*, a local publication in Hudson, New York, what a cocktail was. The editor responded:

"Cock-tail, then, is a stimulating liquor composed of spirits of any kind, sugar, water, and bitters—it is vulgarly called *bitter sling*, and is supposed to be an excellent electioneering potion, inasmuch as it renders the heart stout and bold, at the same time it fuddles the head. It is said also to be of great use to a democratic candidate: because a person, having swallowed a glass of it, is ready to swallow anything else."[28] The editor had essentially described the classic cocktail called the Old-Fashioned, which uses these same ingredients and is still popular today.

By the mid-nineteenth century, the word "cocktail" was growing in popularity. During the Civil War, "Professor" Jerry Thomas authored the first published bartender's guide, which included ten recipes for cocktails. All of these included bitters, as opposed to other mixed drinks in the guide with various names such as slings, flips, toddies, and sours.[29]

In 1917, Mrs. Julius Walsh Jr. of St. Louis, Missouri, held the first cocktail party on record at noon on a Sunday. This was a combination guaranteed to inflame temperance advocates, who were on the rise in the early 1900s. First, Mrs. Walsh's party was clearly about drinking alcohol. It wasn't masked as a dinner party where wine was served with the meal, or a midday meal with sausages, potatoes, and beer. It was about the cocktails.

> *Sometimes food names tell a story, and this may be a good illustration of that concept. The Old-Fashioned is a good approximation of the first cocktail, at least as described in 1806.*

Secondly, the party was held on a Sunday, and drinking on Sunday was particularly abhorrent to the Prohibitionists because they felt it violated the Biblical commandment to keep the Sabbath day holy. (In fact, ice cream sundaes were reputedly developed as a Sunday alternative to alcoholic beverages. Ice cream sundaes, along with temperance-friendly carbonated soda drinks, were served at the new soda fountains, often contained within drugstores. These became popular gathering places for young men and women in small towns and big cities across the country.)[30]

But it was in the Roaring Twenties that cocktails really starred as a symbol for an age, an ironic symbol indeed. Congress ineffectively banned all sales of alcohol during this era in response to years of campaigning by

temperance leaders. In 1919, the same year the states ratified women's right to vote in the Nineteenth Amendment, they also ratified the Eighteenth Amendment, outlawing the sale and manufacture of intoxicating liquors. Congress then passed the enabling legislation via the Volstead Act, which closed every bar and saloon on January 17, 1920. One can only imagine how wild the parties were on the New Year's Eve prior to that ban.

Of course, all Prohibition did was force the liquor trade underground. Organized crime received a tremendous boost, as the racketeers and the bootleggers found new, thirsty customers throughout the nation. Formerly law-abiding citizens found themselves criminalized for home brewing beer or simply enjoying a glass of red wine with their meals. Many old-growth vineyards in California were destroyed or abandoned, stunting America's young wine industry for decades.

IF ALCOHOL WAS BANNED, WHAT MADE THE ROARING TWENTIES SO "ROARING"?

The young flapper with bobbed hair, short skirts, a slim, straight silhouette, and a cocktail in her hand presents the iconic image of the Roaring Twenties, familiar in movies and novels such as F. Scott Fitzgerald's *The Great Gatsby*. But it's not the boozy cocktails that made the 1920s such a rip-roaring time. Many young people in the 1920s experienced a breakout coming of age, enjoying much more freedom economically and socially than their parents' generation had.

New employment opportunities for the increasing number of white-collar jobs lured young adults off the farm, while new technology like the radio and the movies made life more modern and entertaining. The rise of the automobile was particularly transformative. And for the first time, more Americans lived in cities and towns than in rural areas. It was an age of roaring transformation on all fronts.[31]

Without wine and beer, Americans turned to harder substances. Cocktails became increasingly popular at private parties and illicit underground spots called "speakeasies" as Americans started drinking clear spirits, particularly gin, as opposed to the traditional wine and beer or amber bourbon, rum, or scotch. Illegal dark liquor supposedly could be doctored more easily without the consumer seeing the difference. Also—more importantly—private stills could produce gin without worrying about the aging process that darker liquors often required.

It's important to note that the glittering night life and cocktail parties spiked with treacherous bathtub gin were not as pervasive as they may appear to us today, except among a small group of affluent city dwellers.

Bathtub gin, so named because bathtubs were used as vats during the distilling process, varied in taste and effect, since none of its ingredients or processes were regulated. Some clear spirits were actually poisonous. Getting blind drunk took on an even more ominous meaning when people lost their sight after drinking toxic alcohol. More frequently, the gin didn't blind the drinker, but it didn't taste fabulous. So creating a cocktail with nicely flavored mixers provided an easy solution.

Congress eventually repealed the Eighteenth Amendment in 1933, after Franklin D. Roosevelt took office as our thirty-second president. The Great Experiment, as Prohibition was called, had been a failure. Only corruption and organized crime had benefited. And the popularity of cocktails!

Today you can find cocktail menus at various bars and restaurants that offer a lengthy list of mixed drinks, and many of those featured at speakeasies during Prohibition are making comebacks.

Interestingly, today's "classic" cocktails like gin and tonics, Tom Collins, and especially martinis were all drinks that had been invented before the 1920s but found new acceptance during and after Prohibition despite, or perhaps because of, their illegality.

GIMLET

The first gimlet, a gin cocktail made with Rose's lime juice, made its appearance in 1928, and a good bartender can make one today in the blink of an eye. A gimlet is a hand tool with a sharp piercing point, used for drilling small holes, and a judgmental individual is said to have a "gimlet-eyed" stare if he or she disapproves of some action or expression. So the gimlet drink should be sharp, not sweet, although hopefully it won't drill holes in your head.

INGREDIENTS

- 1½ ounces gin
- 1 ounce Rose's lime juice
- 1 generous twist of lime
- ¾ cup cracked ice

DIRECTIONS

Place the gin, lime juice, lime twist, and ice in a cocktail shaker and shake vigorously. Strain into a glass and serve.

Makes 1 gimlet cocktail.

BITE 69

CANAPÉS

Passed Hors d'Oeuvres

The rise of cocktail parties also made hors d'oeuvres newly popular, although little bits of food before a big meal have appealed to diners for a long time. In fact, people have consumed appetizers for millennia. The Romans ate small bits of tempting food, such as olives, or even stuffed dormice, before their banquets began. Hors d'óeuvres, a French term meaning "outside the main work" of the meal, appeared as small plates surrounding the main dishes on a formally draped table, a dining fashion emulated in other western European countries as well as Britain and America until the mid-nineteenth century.

By the late 1800s, the style of dining service in elite homes and restaurants changed from placing many types of food on the table at one time to a series of successive courses. The hors d'oeuvre course during the Gilded Age, for example, was served after the soup course but before the fish course. These three courses might be followed by five or six more, creating a formal nine-course meal.[32]

It took the rise of the cocktail party to move hors d'oeuvres off the formal table at an elegant, lengthy dinner to become tempting tidbits passed on a tray. Small appetizers passed on trays are also referred to as "canapés." Canapé comes from the French word for sofa, perhaps because different tasty things sat on top of the toast or cracker. Strictly speaking, canapés were various composed toppings that covered crackers or a thin, petite round of bread, such as a cucumber with dill on top of a buttered round or, more luxuriously, fois gras and a slice of seedless grape on a toasted bit of brioche.

> Cocktail parties broke through the rigidity of Victorian-style receptions and introduced a lively atmosphere, sometimes with dancing.

The hosts or waiters passed canapés on trays at private cocktail parties and receptions, while heavier platters sat on a table, waiting for guests to help themselves. This style of entertaining meant that people were free to stroll about while sipping a beverage of their choice, eat what they wanted, and talk to whomever they pleased. It felt festive and modern to the upper-middle classes who were raised with a more formal entertaining style. Jazz and popular music played on the Victrola record player. No one sat for very long any more, and no one was stuck with a dull-as-dust dinner partner for nine courses. Stuffy, hierarchical seating did not die for formal, elite events, but at least there were alternatives.

Long after the end of Prohibition in 1933, the traditions of cocktail parties and passed hors d'oeuvres have continued to thrive in America. The music has changed, and so have the fashions (even what canapés are served), but the general aspects remain the same. People complain about the light food (When's real dinner?), the incessant standing (When can I sit down?), or the excess or lack of alcohol (I think I was over- or underserved), but it continues to be a relatively easy way to gather a crowd of people together and have a good time.

BITE 70

Tostadas

Tortilla chips (or tostadas) frequently make an appearance at cocktail parties today, or at simple gatherings around the television. Prior to the 1920s, however, they were always handmade, until a man named José Bartolomé Martinez mechanized the process.

To the majority of Americans in the early twentieth century, the mechanization of food processes was a godsend. For more than ten thousand years, women and domestic workers of both genders had labored to prepare everything from simple daily meals to elaborate feasts for their families, clans, employers, and owners. Today we put enormous value on the phrases "homemade," "handmade," and "old-fashioned." But the technological contributions of inventors to improving the standard of living for those who spent time preparing meals should not be underestimated.

In the remote areas of the American Southwest, well into the twentieth century, Mexican Americans and American Indians continued to grind their maize by hand. In cities and towns, however, mechanized mills replaced that slow process, producing good-quality masa (corn dough) and corn tortillas that were fresh and cheap.

José Bartolomé Martinez ran Azteca Mills in San Antonio, dominating the local market with his own patented Tamalina corn flour, fresh tortillas, and masa. In 1912, he started using his leftover masa to create perhaps the first commercially produced corn tortilla chips, or "tostadas." He sold them in eight-ounce waxed paper bags bearing his logo of an eagle holding an ear of corn.[33]

Azteca Mills marketed the chips to restaurants by shaping them in small triangles, which made them easy to use with dips. The machine-made tostadas were inexpensive, tasty, and immediately popular. From Mexican-themed restaurants, they spread to private homes. Families had made corn chips in their own kitchens, but it was a messy, time-consuming process. Martinez's chips brought machine-made snack food into homes around the region.

Like many great inventions, Martinez's mechanical process was a simple one. In this case, it was also easily copied. Several small competitors started selling the chips. Gustavo Olquin, originally from Oaxaca, started his own chips business in San Antonio but longed to return home. C. Elmer Doolin, a young man who hailed from Kansas but had moved to San Antonio looking for new opportunities, bought Olquin's entire business and his recipe for one hundred dollars in 1932.

Doolin named the chips "Fritos." He relocated the business to Dallas, then built another plant in Tulsa, and another in California later in the decade. By the 1950s, Doolin's Fritos became the brand that meant corn chip to Anglo-Americans.[34] He went on to develop Cheetos and, perhaps more importantly, a bag rack for stores to display bagged snack food, which is what we still see in grocery stores today. His company eventually became Frito-Lay, and Lay's potato chips joined the product line.

Then, in 1965, Frito-Lay merged with Pepsi Cola, forming PepsiCo, one of the world's largest food and beverage conglomerates today. Its stunning headquarters in Purchase, New York, are a long way from old San Antonio, and it's hard to believe PepsiCo's roots reach back to Texas in the 1920s. But that era's influence is still being felt in many ways, including in our sandwiches, our cocktails, and our choice of snacks.

So many changes in Americans' diet and attitudes toward food began in the first part of the twentieth century. Nationally branded, packaged convenience foods appeared at grocery stores across the country. Scientists discovered vitamins and other nutrients, and understood for the first time the scientific link between dietary deficiencies and disease.[35] As more and more people moved off the farms and into the cities, our understanding of the food on our plate, and the way we related to what we ate, changed forever.

SCRAMBLED EGGS, HERSHEY BARS, AND PEACH COBBLER

The Great Depression and World War II

he Great Depression and World War II are the two cataclysmic events that serve as the central turning points of the twentieth century around the globe. They changed the course of history, and they changed the lives of an entire generation that has become known as the Greatest Generation in America. (The name is based on the title of a bestselling history, *The Greatest Generation*, by journalist Tom Brokaw.) Although the United States had suffered through many boom-and-bust cycles, nothing matched the economic collapse that lasted almost a decade. And World War I became the prelude to a bigger, more massive, and more brutal conflict than anyone could dream.

Throughout the Depression years, men and women lived their lives as best they could. Many people prospered, but others foundered, while most tightened their belts. Some left the home where their families had lived for generations, looking for opportunity wherever they could find it. And when the war came in 1941, Americans from all backgrounds rolled up their sleeves to help with the war effort, on the home front or in the service.

BITE 71

❧ MULATTO RICE ❧
The Harlem Renaissance

Too often we peg the beginning of the Great Depression to the sudden stock market crash in 1929. But farmers in many parts of the country had already felt the pinch—that soon turned into a choke hold—after the Versailles Treaty was signed in 1919, formally ending World War I.

The Roaring Twenties brought unheard of prosperity in terms of wages and consumer goods to many middle-class white Americans, especially those living in urban areas. Simultaneously, it brought devastation to other parts of the country, especially as a catastrophic agricultural depression settled over the South and parts of the Midwest.

World War I had created a vast, international market with high prices for American farm commodities. It was a boon for farmers unlike anything they'd ever seen before. With peacetime came a rapid decline in demand, followed by abysmal weather conditions including the infamous Dust Bowl, caused in part by a series of ruinous storms in the early 1930s. Farm families were dislocated—20 percent of all family farms were sold to pay debts or stave off financial ruin between 1920 and 1932. For them, the Great Depression did not begin with the stock market crash on Black Tuesday, October 29, 1929. It began in 1920, when agricultural prices fell through the floor.[1]

Black Americans, especially the vast majority who lived in the rural southern states, were particularly vulnerable in this era of economic desperation. The Jim Crow laws, the lynchings, and the sporadic, race-based attacks that had terrorized African Americans, especially after the Civil War, continued unabated after World War I, while membership in the Ku Klux Klan grew to unprecedented heights. Young African American men who had fought with distinction during the Great War returned from the Armistice to bitter prejudice in all parts of the country. But in the South the hatred was particularly virulent. The cotton market had plunged, and share-cropping was just a few rungs above chattel slavery. For many black people, it was time to look for a better life.

So the Great Migration north, which began with a trickle of people of color traveling out of the South in 1900, turned into a full-on flood in the 1920s. African Americans left the southern regions where they had lived, some for generations, to find higher wages and more respect. Cities such as New York particularly benefited from this influx of talented young people eager for new opportunities. The Great Migration transformed the northern Manhattan neighborhood of Harlem, long ago a Dutch enclave.

Once a rural suburb of Manhattan, Harlem was where wealthy New Yorkers had spent their summers. As the neighborhood became more densely populated, new immigrants moved in. Harlem was one of the few places in Manhattan where black people could rent apartments, though the landlords jacked up the prices for their new clientele. Still, Harlem began to thrive with a burgeoning art, letters, and music scene, along with a world of newly transplanted professionals.

It was the era of the Harlem Renaissance in the late 1920s and '30s. Young authors like Langston Hughes, Richard Wright, and Zora Neale Hurston brought new voices to the page, while jazz artists like Duke Ellington and Lena Horne brought an innovative sound. Clubs and speakeasies blossomed like colorful, night-blooming flowers. The most famous one of these, the Cotton Club, offered a pseudo-exotic experience for elite whites, with live music and performers often depicting African jungle fantasies. Run by a gangster named Owney Madden, the club allowed people of color in the back door, but only as performers or staff. Duke Ellington finally convinced Madden to integrate the Cotton Club, which made it even more popular.[2] Less famous watering holes also welcomed an integrated crowd, where live music and bootlegged liquor attracted all sorts of clientele.

Street vendors selling traditional southern foods brought the regional aromas of the Cotton Belt to New York sidewalks. Hot corn, pigs' trotters, fried chicken, sweet potatoes, and watermelon slices created a street-side banquet for adventurous tourists or locals on their way home from work. New Yorkers, white and black, were delighted to try these new delicacies while walking down the street. All of Manhattan, including Harlem, was on the go.

Immigrants from the West Indies joined the transplants from Dixie, adding their lilting patter and selling exotic imported bananas, coconuts, and pineapples alongside the traditional southern fare. Small-scale entrepreneurs, colorful characters on the city streets, could make a good living if they carefully managed their costs.

In *High on the Hog*, author Jessica B. Harris tells of "Pig's Foot Mary," who died in 1929, having amassed $375,000—a veritable fortune in those days—selling pigs' trotters, corn, and chicken from her cart. Another opportunistic entrepreneur, Patsy Randolph, collected watermelon rinds for free from fruit stands, pickled them, and sold them in glass jars. The watermelon pickles brought memories of the South to the mouths of her clients, who loved the opportunity of Harlem but missed the flavors of down home.[3]

People went informally into the food business by holding rent parties where, in an attempt to pay the high rent, a group of roommates might turn their flat into a catered club for one night. Once again, southern food played a starring role. Paying guests lined up for big pans of fried chicken, ribs, corn bread, collards, and homemade cake. You also might find a delicious, sustaining tomato-and-rice dish called "mulatto rice" on the table.

The old word "mulatto" refers to someone of mixed race ancestry, usually half black and half white. Today the term is sometimes seen as offensive or wrong-sounding because it hearkens back to a day when everyone was defined in terms of race, by law and by cultural practice. But in the 1920s and '30s, people used the term mulatto frequently. I don't know how the rice and tomato dish got its name, but Zora Neale Hurston mentions it in *Their Eyes Were Watching God*.[4]

The following recipe excerpt appeared in *The Savannah Cook Book*, by Harriet Ross Colquitt. Published in 1933, the cookbook reflects the pervasive (and to modern eyes, appalling) racism that was part of white culture, particularly in the American South.[5] This recipe is no exception, and while Colquitt's first line may spark incredulity, it is historically accurate and reflective of the time.

While the language of the recipe is offensive today, the rice is a delicious part of our national cuisine. Perhaps it deserves a race-blind name, like Savannah Tomato Rice, in recognition of the rich southern heritage of so many great American dishes. Or perhaps we should keep the name as a reminder that people of mixed race contributed particularly tasty flavors to the American plate, even as they faced racism on a daily basis.

MULATTO RICE

Harriet Ross Colquitt introduced this recipe with the snooty and overtly racist comment, "This is the very Chic name given to rice with the touch of the tar brush."

Fry squares of breakfast bacon and remove from the pan. Then brown some minced onion (one small one) in this grease, and add one pint can of tomatoes. When thoroughly hot, add a pint of rice to this mixture, and cook very slowly until the rice is done. Or, if you are in a hurry, cold rice may be substituted, and all warmed thoroughly together.[6]

Serves 8 as a side dish.

I would add 3 cups of water or chicken broth to the onion and tomato mixture, which I think the original author left out by mistake. If you use leftover cooked rice, leave out this extra liquid. Also, I would add the crumbled bacon and 2 tablespoons freshly minced parsley to the rice before serving.

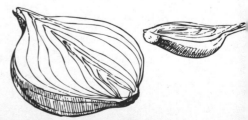

BITE 72
WPA SOUP
Soup Kitchens in the Great Depression

Like rice, soup is a versatile food. It ranges from the most luxurious to the most basic and can be worthy of divinity in either incarnation. For centuries, religious groups, governments, and private individuals have served soup to the needy, because it can be very economical and nourishing as well. It's also one of the easiest foods to produce in large quantities, and best of all, expanding the number of servings is relatively easy, even after the cook has embarked on a certain amount. From medieval Europe, when the harvest failed, to Ireland in the Great Famine of 1847 to the United States in the Great Depression to our own soup kitchens today, men and women have ladled out soup to long lines of people of all ages. And we nourish ourselves with it too, and rightfully so.

The Great Depression supposedly began with Wall Street's crash in 1929. Like all stunning financial crashes, however, many signs of economic weakness predated Black Tuesday. Misguided protective tariffs hampered international trade. Agriculture had never recovered from overproduction and the ensuing price deflation after World War I. Frenzied speculation, along with unregulated and risky banking practices, culminated in a financial debacle that would take years from which to recover, ultimately requiring the buildup to World War II.[7]

This was not the first time that Americans had experienced an economic depression. Nor was it the first time that people went hungry and lined up at soup kitchens. It was, however, a larger, more catastrophic economic event, affecting more people from more places than ever before, and that's where soup began to play a vital role.

A few days after the stock market tanked, Franciscan monks and community volunteers set up a soup kitchen in Detroit. They served between 1,500 and 3,000 of the hungry every day.[8] Soon soup kitchens sprouted up in major cities and small towns as the number of destitute families grew. Sometimes state and local governments contributed to these volunteer efforts, though often where the need was greatest, the relief was the least. The soup was frequently thin and meatless, but it was hot and usually accompanied by bread, so it was a godsend to anyone who was hungry. The bread, often stale, added important calories and some nutrition when soaked in the hot broth.

President Herbert Hoover, who had run the U.S. Food Administration effectively during the First World War, urged the creation of soup kitchens. He believed that local volunteer efforts would work as well in 1929 as they did in 1917. But it would take more than free soup to pull America out of the biggest economic collapse it had ever experienced. Although Hoover eventually funded some programs, his efforts were too little, too late.

The election of Franklin Delano Roosevelt in 1932 ushered in the New Deal, in which the federal government poured money into a wide variety of jobs programs. The New Deal was not universally popular but it did provide new opportunity and hope for many. During Roosevelt's first administration, Congress enacted legislation for the Works Progress Administration (WPA), creating government-funded jobs to help "put America back to

A line for soup and other comestibles offered to the hungry and unemployed during the Great Depression.

work." Work of all kinds, from food writing to road paving, was federally funded.[9] The New Deal did not end the Depression completely, but it had a positive impact on a withered economy, renewing many people's faith in themselves and in the federal government.

Today there are still homeless and hungry people in America. The majority, sadly, are children. The professionals and volunteers who work at local soup kitchens and food pantries do their best to fill the empty stomachs of America's neediest individuals, young, old, and in-between.

WPA SOUP
(Potato Sausage Soup)

This soup received its name because the WPA provided the necessary funds to pay for the sausage.[10] This should serve about 125 people with a bowl each, or 150 with a cup each.

INGREDIENTS

- 28 quarts (7 gallons) of water
- 4 quarts potatoes, scrubbed and chopped
- 2 to 4 quarts carrots, scraped and chopped
- 2 quarts celery, diced
- 2 quarts onion, diced
- 16 pounds pork sausage, browned and chopped if link, or just browned if bulk. Use money from the government for the sausage and get as much as you can.
- Salt and freshly ground black pepper to taste

DIRECTIONS

Divide the water, potatoes, carrots, celery, and onion into two 6-gallon vats. Bring each to a boil. Reduce to a simmer and cover. Add more water as needed to keep the potatoes from sticking on the bottom and scorching, since those large vats are often thin metal. Cook until the potatoes are fork tender. Add the sausage and continue simmering until meat is done.

Makes 10 gallons.

It's good to let this sit in a cold place for a few hours or overnight before serving, so you can skim the pork fat off the top. If you have access to fresh or dried herbs, add these at the beginning but mince them first or tie in bundles with twine: rosemary, thyme, sage, oregano. Any combination will be okay, but don't let it overpower the soup. Use about 1 tablespoon per gallon if dried.

BITE 73

Lamb's Quarters
Foraging for Food

Soup couldn't fill everyone's stomachs in this era of need, however. As mentioned earlier, farming families faced particular hardships during the Great Depression. Many lost their land to foreclosure, unable to pay their mortgages when the price of agricultural produce fell to rock bottom. To add to the farmer's woes, Mother Nature stepped in with the worst drought on record at the time, from 1933 to 1935. Decades of poor farming practices that had led to soil erosion only worsened the nightmare. These were the Dust Bowl years.[11]

Most Americans had grown accustomed to a plentiful diet. Farmwives were proud of their heavily laden tables, groaning with the weight of food from their pastures, fields, and gardens. But in the 1920s and 1930s, swaths of America's heartland became a place of dust and drought, a veritable food desert. In other areas, productive harvests lead to the massive slaughter of farm animals and destruction of crops to prop up farm prices in 1933, even while people starved in neighboring counties.[12]

Old World recipes designed to stave off hunger at little cost resurfaced in the kitchen. Boiled potatoes seasoned with melted lard and chives made a filling meal. Cornmeal mush, that colonial standby, reappeared as a cheap antidote to hunger pains.

Foraging also added valuable nutrients to the family table. Children gathered wild berries and collected tender, young dandelion greens, packed with vitamins, after a long winter of no fresh food. Fiddleheads from the forest floor and common weeds like silvery lamb's quarters and spinach-like pigweed also provided fresh greens when the spring garden had turned to dust.

A now-elderly woman from Menomonie, Wisconsin, remembered growing up on her family's farm during the Great Depression. Her mother raised chickens and traded eggs at the local store to feed her family. Her mother told her that "one of her lowest points was when she had to can weeds. The garden didn't do well at all during the driest time, but the weeds, lamb's quarters, still flourished, so she canned them for something on the cellar shelves against the winter coming on. We must have eaten a lot of lamb's quarters in those hard years."[13]

Today, folks interested in slow food and traditional foodways still find foraging particularly rewarding. Wild leeks known as ramps find their way into many gourmet dishes at high-end restaurants at the first sign of spring. Dandelion greens and other plants known as weeds to most suburban lawn owners are also appearing on menus and in dishes nationwide, as we look to return to reaping the bounty from the land beneath our feet. Personally, I adore fiddleheads, dark forest green, sautéed in olive oil, and sprinkled with toasted walnuts. I have to buy those at a farmers market. More and more people see these wild things as a wonderful way to expand our gastronomic experience and eat completely fresh organic food.

Long forgotten today is the fierce desperation of canning weeds to nourish your hungry children and feed the gaunt men as they gather round the table during a cold Wisconsin winter and look at you expectantly, hoping that somehow you'll produce a bigger, more satisfying meal. The hardship of those lean years of the Great Depression lasted far too long, especially for the farm families in the Dust Bowl states. It would take the increased government spending and ultimately World War II to bring full prosperity back to the United States.

BITE 74

⚜ Eleanor Roosevelt's ⚜ Scrambled Eggs

The 1932 election of Franklin Delano Roosevelt as president brought more than just hope and the promise of an activist federal government. It also brought his progressive wife, Eleanor, a smart, cultured, "modern" women with a deep belief in human rights.

Eleanor and Franklin Roosevelt's marriage was a partnership that benefited the United States and gave each of them the platform they needed to achieve their goals. Some of these goals were shared. President Roosevelt's infidelities hurt Eleanor immeasurably but she chose to remain his wife, knowing that divorce would kill his career. Perhaps Eleanor also recognized that being First Lady allowed her to work throughout the country on civil rights and issues of poverty that she cared about deeply. Being First Lady also insured national press coverage for her several causes.

Mrs. Roosevelt was an accomplished writer, speaker, and campaigner who acted as her husband's eyes and ears as she traveled tirelessly around the nation during the Great Depression. She was not, however, a good cook, and she had no interest in the White House kitchen.

In fact, food at the Roosevelt White House was notoriously lousy, except for the bread. Eleanor had hired a loyal supporter, Henrietta Nesbitt, who was a good baker but incapable of putting a decent meal on the table—or planning one, for that matter. Claiming that fine food was inappropriate during the severe economic crisis, Mrs. Roosevelt did not care about complaints from guests or her husband's

> *Mrs. Roosevelt had been allocated household management as one of her First Lady responsibilities and, perhaps in annoyance at a task she resented, refused to get rid of the dreadful cook, Mrs. Nesbitt. Like many women of her wealthy class in that era, Eleanor did not know how to cook, either.*

wishes. Perhaps the simple loyalty of the brusque Mrs. Nesbitt, who managed to alienate long-term members of the household staff, compensated for her lack of culinary talents from the First Lady's point of view.[14]

Mrs. Roosevelt did make one dish every week, however, at which she was quite adept: scrambled eggs. She and her husband hosted small, casual Sunday suppers for their advisers and intellectual guests. President Roosevelt manned the bar from his chair, as his polio reduced his mobility considerably, and produced shakers of martinis or whiskey Old-Fashioneds spiked with orange juice. (His wife, not much of a drinker, might sip a sherry.) The staff put out plates of cold meats and salads. And Eleanor stirred scrambled eggs in a silver chafing dish for everyone. Apparently, they were tasty too.

NOT EVERYONE WAS BORN TO COOK...

There are significant class overtones to Eleanor's lack of cooking ability. She was raised with a cook, and she employed a cook. Upper-class and upper-middle-class women of her era often did not know a thing about cooking a meal and were proud of it. In fact, they frequently boasted about their inability to boil water.

But there was something about scrambling eggs that meant you were independent and could feed yourself. While "roughing it" at their summer homes before the cook arrived, for example, women of a certain class could always manage the rather simple culinary skill of scrambling eggs in a black, long-handled frying pan over an open fire. It represented just the right combination of can-do spirit and lack of interest in the details of domestic service.

These were relaxed, chatty suppers, with guests well lubricated by the cocktails created by the nation's mixologist in chief, the president himself. With limited protocol and loosened tongues, advisers and guests spoke freely, sharing their insights and observances. President Roosevelt was wise enough to value the brilliance of these intellectual people's work. Sunday suppers at the White House became a relaxed time during a tense age, a chance to garner the latest ideas from some of the great progressive thinkers during the New Deal. Members of the White House staff called these informal dinners "Scrambled Eggs with Brains."[15]

BITE 75

SPAM

In 1937, during Franklin and Eleanor Roosevelt's stay in the White House, the Hormel Foods Corporation of Austin, Minnesota, introduced a new product: SPAM, a canned, precooked pork shoulder and ham product. The name derives from its contents: **S**houlder of **P**ork and h**AM**.

Hormel's timing was perfect. The economy was still struggling, and people appreciated inexpensive canned meat, which was far more affordable than buying the fresh version at the butcher's. Four years later, when the United States joined World War II, SPAM became a ubiquitous part of the American diet, for civilians as well as military personnel.

With the rationing of fresh meat during the war years, canned SPAM provided a cheap, protein-rich alternative along with the convenience of an indefinite shelf life—a shelf life so long that horrified foodies today consider it comparable to that of uranium. Many soldiers, especially those who were shipped overseas early in the conflict, thought SPAM was unique to the armed forces and nicknamed it Special Army Meat. But their families back home were often eating the same processed pork. Fried, sliced SPAM, SPAM and potato casseroles, spaghetti with tomato and SPAM sauce: the uses of SPAM were multitude.

Civilians in other countries learned to love SPAM as well. The U.S. Navy introduced SPAM to the local inhabitants as they wrestled island after island from the Japanese as war raged across the Pacific. The canned meat remains hugely popular in Korea, Guam, Okinawa, the Philippines, and

Hawaii, where residents consume more SPAM than in any other state. In the United Kingdom, home economists graced their readers with reams of recipes starring the canned meat during the long years of meat rationing that continued until 1954, nine years after the war ended.[16]

The story of SPAM takes us to one of America's most unjust actions during World War II. In twentieth-century history, World War II has an overarching theme of heroic sacrifice for democracy and freedom. But like any war, its story is more complex. In early 1942, FDR signed the executive order that lead to the relocation of 122,000 Japanese Americans to internment camps in remote areas of the western states. Families were forced to abandon their homes, possessions, and pets to live at these sites, also known as relocation or concentration camps.

Fried SPAM became something of a comfort food for some Britons, and a punch line for many others. The classic Monty Python skit comes to mind here. Apparently, the Monty Python SPAM café inspired the term "spam" meaning electronic junk mail.[17]

They lived in spartan barracks, surrounded by barbed wire, and ate in communal dining halls. The food proved to be one of the hardest adjustments for many of the internees. Many meals consisted of potatoes and processed meat—hot dog "weenies" or SPAM. Interestingly enough, the internees disliked the potatoes, considering them the most distasteful part of their rations.

They missed rice, the core ingredient of the Japanese diet. But they consumed the SPAM and weenies, adapting them into Japanese American food traditions after the nightmare of internment ended. Today, the continued popularity of Spam Nori, a SPAM-based sushi, illustrates how food cultures merge (like Mulatto Rice), even when injustice forced the juxtaposition unfairly upon a part of the population.[18]

BITE 76
MEATLOAF

World War II was the first time the federal government mandated food rationing. The Meatless Mondays and Wheatless Wednesdays of the Great War had asked for voluntary participation during our relatively brief engagement overseas between 1917 and 1918. But the massive mobilization and civilian support required by World War II meant that rationing was more than a cheerful patriotic act. It became the law of the land.

The Great Depression had forced households to cut back on food expenses, so reducing meat was nothing new. Food reduction required by financial necessity can leave a bitter taste, but rationing on the home front enabled all civilians to do their part for victory and thus was much more positively accepted.

In May 1942, six months after the Japanese bombed Pearl Harbor and the United States once again joined the Allies in a global conflict, the U.S. Office of Price Administration (OPA) froze prices on practically all everyday goods. Every household received war ration books and tokens that dictated how much meat, gasoline, nylons, shoes, and tires could be purchased by any one person. Planning meals, and using recipes that called for a limited amount of rationed ingredients, required civilians to think carefully about their food consumption.

> *The war consolidated meatloaf's "high-ranking position in the housewife's culinary artillery," according to food writer Nadia Arumgum.[19]*

Once again, the government encouraged the planting of Victory Gardens and home canning. Ideas for meat extenders flourished. Eating meatloaf also became an all-American duty.[20]

Meatloaf appeared in cookbooks soon after the invention of mechanized meat grinders for home kitchens, around the turn of the century. Meat grinders meant the toughest cuts of beef—also the cheapest kind—could be incorporated into a variety of dishes. Through World War I and the Great

Depression, ground meat extended with some sort of starch and cooked in a loaf pan grew in popularity. By World War II, it was a beloved dish, served at home and at homey diners.

The OPA ended meat rationing in November 1945, just a few months after the Japanese surrendered. But unlike some (pretty dreadful) war-ration recipes, which never saw the light of day again once peace returned, meatloaf remains as one of America's most beloved home-style foods.

ROSIE THE RIVETER ATE MEATLOAF

Rosie the Riveter symbolized the women workers newly welcomed into the defense industries during World War II to replace the men enlisted abroad. No doubt, she would have brought meatloaf sandwiches for lunch at the munitions factory. Between 1940 and 1945, with massive numbers of enlisted men leaving industrial jobs vacant, the number of women workers rose by 50 percent, from twelve million to eighteen million. Civilian women became railroad workers, streetcar conductors, and engineers. They worked in steel mills and other heavy industries. My mother drove an oil truck in Chicago, and then worked in a troop transport. I bet she ate meatloaf too.[21]

BITE 77

⚜ HERSHEY BARS ⚜

We Americans *love* our chocolate, and we're not the only ones. Chocolate, made from the cacao beans native to South and Mesoamerica, has a long and fascinating history. Archaeologists have found ceramic vessels with traces of chocolate in them dating from around 1700 BC at an ancient Olmec site near Veracruz, Mexico. Two thousand years later, the Mayans boasted a variety of recipes for serving chocolate hot or cool, as a medicine, or with added cacao butter.

After being introduced by the Aztecs to the wonders of chocolate during their first encounters on Columbus's voyages, the Spanish brought chocolate home but considered it such a powerful aphrodisiac that they kept the recipe locked up in a monastery for most of the 1500s. Eventually, word got out, and thick, hot chocolate, now sweetened, took the aristocracy by storm in the seventeenth and eighteenth century. Like coffee and tea, hot chocolate developed its own luxury consumer items, with special pots, cups, and cooking utensils designed to enhance the delectable experience of conspicuous consumption. Europeans also began to eat chocolate as a candy and as a dessert flavoring for the first time during this period.[22]

In the 1820s and '30s, the Van Hooten family in the Netherlands developed a method of processing cacao beans that is still known today as "dutch-processed." This technique created a high-quality but reasonably priced cocoa powder, which is the basis of much of the world's chocolate today. By 1875, milk chocolate candy had been invented in Switzerland.

Milton Hershey was a young American caramel manufacturer who traveled to Chicago's World Columbian Exposition in 1893. There, he became convinced that the future of candy making lay in chocolate, after seeing the confection being made with the latest technology. By 1900, he had developed the Hershey Bar, a milk chocolate bar that Americans embraced. He sold his caramel business for the then-stunning sum of $1 million and, in 1903, began construction of a chocolate plant in his hometown of Derry Church, Pennsylvania—soon to be rechristened "Hershey."

Our modern word "chocolate" is probably based on the Aztec (Nahuatl) word xocolatl, *meaning bitter water, for the Aztecs introduced this beverage to Christopher Columbus in 1502. Chocolate was drunk exclusively by Aztec royalty and warriors, who consumed this sacred drink mixed with different flavorings, like hot chili peppers. Cacao beans, the basis of all chocolate, were so valuable that they served as a type of currency. (In fact archaeologists have discovered counterfeit cacao beans dating from the Aztec period.)*

In 1907, a flat-bottomed, teardrop-shaped piece of milk chocolate, called the Hershey Kiss, hit the marketplace. Individually wrapped in foil by hand, the kisses were bite-sized and affordable. By 1921, a machine wrapped each kiss in foil and added the little paper ribbon bearing the name Hershey, which is still part of each Hershey Kiss today.

Other chocolate products followed, including Mr. Goodbar (1925), Hershey's Syrup (1926), and Krackel (1938).

But it was the military Hershey Bar, known as the D Ration, that made history during World War II. In 1937, army quartermaster Colonel Paul Logan approached the Hershey Company about making an energy bar for the military. He had four requirements for the bar:

1. Weigh four ounces
2. Be high in food energy value
3. Be able to withstand high temperatures
4. Taste "a little better than a boiled potato" (so the soldiers would reserve the bars for emergencies and not use them as a sweet snack)

The D Ration proved successful in tests and fulfilled Colonel Logan's fourth military requirement by using oat flour as an ingredient, keeping the military's version of chocolate from being too tempting to servicemen. When the United States entered World War II in 1941 after the Japanese bombing of Pearl Harbor, the Department of War ordered massive shipments of the bars, which practically required soldiers to develop strong teeth and weak taste buds. In response to government requests, Hershey also invented the waxy Tropical Bar, which traveled well in the scorching heat of the South Pacific and tasted a bit better. Either way, soldiers often gave these less-than-delicious bars to civilians, especially children, who were delighted by the gift. Young recipients often needed the calories desperately and would remember American GIs very fondly for their generosity during wartime.

Today, eighty million Hershey Kisses are produced each day around the world.

Between 1940 and 1945, more than three billion D Ration and Tropical Bars were produced and distributed throughout the world. Hershey

Chocolate Company was awarded the Army-Navy E Award for excellence in exceeding expectations for quality and quantity in the manufacturing of these products, although the troops often detested the taste.[23]

In 1945, when the troops finally came home, Hershey Bars made of real chocolate were waiting for them. And they've been incredibly popular since.

BITE 78

❧ PEACH COBBLER ↝

Foods rich in memory and meaning waited on the home front for most the returning military personnel at the end of the war. The demobilization of troops in 1945 and 1946 meant that millions of men—and 350,000 women— were coming home. And America was eager to welcome them with open arms and heavily laden tables. Though rationing remained in effect for more than a year after the Japanese surrender in August 1945, families, friends, and neighbors managed to create bountiful meals for joyous reunions with their loved ones returning from war. Generally, mothers, wives, and other women prepared the new veterans' favorite dishes, as at-home food preparation was even more gender-based than it is today. These welcome-home dinners often represented the returnees' culinary heritage and background.

For example, a sailor returning home to Atlanta in 1946 might enjoy a welcome meal, prepared by his mom and sisters, of sweet iced tea, lemon-carrot Jell-O salad, baked ham with a Coca-Cola glaze, hot biscuits, and a delicious, warm peach cobbler for dessert. A classic baked fruit dish, cobbler plays a humble though beloved role on the traditional dessert table, particularly in the American South.

Like a double-crust fruit pie, it has a top, a bottom, and a sweet filling, but it also resembles fruit crisps, brown bettys, pandowdies, or crumbles by being baked and presented in a rectangular pan, not a circle. And not all cobblers had a bottom crust, which made them easier to cook in an open hearth during the colonial period. Food historians credit Lettice Bryan with publishing the first written description of a cobbler in her 1839 cookbook *The Kentucky Housewife*. Normally considered a family-style dessert, cobblers were tasty enough for company, even if the topping might look more casual than some fancy pastries.

Southerners especially loved peach cobblers because peach trees naturally thrive in their warmer climate. First cultivated by the Chinese around 500 BC, peaches traveled to the New World with the Spanish explorers, escaping orchards to grow wild in the South. Imagine the lucky pioneer family in the early 1800s, stumbling upon a small grove of wild peach trees in the forests of what is now Georgia. What a delightful surprise![24]

The family-based welcome-home meal took place among all ethnic and racial groups, if the veteran had a family or loved ones to welcome him home. Of course the menu would vary according to the region, personal tastes, and culinary heritage of the individuals.

Among some groups like Italian and Japanese Americans, their cuisine closely reflected their cultural identity. Although Japanese Americans had been forced into concentration camps for much of the war (as noted in Bite 75), their sons of military age served with extraordinary honor on the European front. Enlisted black American men signed up to fight for their country, and they too faced enormous prejudice, often relegated to non-combat positions as stevedores and maintenance men. Once again, the outstanding records of African American fighters, such as the Tuskegee Airmen and the 761st Tank Battalion, were not enough to change hardened racial attitudes in the military and in broader American society.

Many African Americans supported the Double V, a campaign for two victories—victory over the Axis powers overseas and victory over Jim Crow laws at home. It seemed like a reasonable premise: if a man was willing to sacrifice his life for his country, his country should be willing to treat him equally and fairly regardless of his color. Particularly in the southern states, that was far from an accepted truism. There was a lot of hard work left to do to achieve victory at home over racial injustice.[25]

However, in 1945 and 1946, for black men coming home (and for servicemen of all colors and creeds), the first family gathering was a time to be with your own people, eating, catching up, and relaxing. The women would weigh the table down with whatever dishes they knew would delight, maybe splurging a little on ingredients. One of my favorite cookbook authors, Edna Lewis, describes a celebratory meal in Freetown, Virginia, served in late summer. Dessert was peach cobbler. The meal offers southern food, and it sounds like an ideal welcome-home dinner for a grown-up son, lean, weary, and happy to be back from the battlefields of World War II.[26]

There would be another battle waiting, this time for equality in America. Ultimately, the Double V campaign led to the civil rights movement of the 1960s. But for now, this World War II victory was sweet, and it was good just to be home.

MENU: WELCOME-HOME SUPPER

This menu is inspired by Edna Lewis, one of the great American cooks of the twentieth century whose cookbooks are great literature. Lewis was a southerner from Virginia who introduced the rest of the world to her magical regional specialties with her first of four books, *The Taste of Country Cooking*, in 1976. The descendant of slaves, she preserved a delicious gastronomical heritage, sharing it eloquently with her readers through her recipes, menus, and memories. A soldier from Lewis's community, returning to Virginia from the battlefields of Europe, might be welcomed home with a meal like this.

WELCOME-HOME SUPPER

- Boiled Virginia Ham
- Hot Shredded New Cabbage
- Smoked-Pork-Flavored Green Beans
- Glazed Beets
- Watermelon Rind Pickles
- Biscuits with Butter
- Fresh Peach Cobbler, still warm from the oven, served with heavy cream spiced with nutmeg

❧ NAVAJO FRY BREAD ❧

As travelers and immigrants can testify, taste memories can be very powerful, especially when a long distance, either geographic or temporal, separates a person from that particular food. It seems likely that servicemen overseas sorely missed some types of the food they loved at home, whether that food was handmade or came in a can. The Navajo code talkers in the Pacific Theater during World War II probably missed fry bread.

With the outbreak of hostilities in 1945, the Department of War desperately needed men and materiel. They also required secret codes for military communications. Brilliant cryptographers labored to develop undecipherable languages while inventors sweated over technological devices, all aimed at keeping information secret from the Axis powers—Germany, Italy, and Japan. But it was incredibly difficult to keep these messages encoded and to ensure the enemy didn't intercept or break the code.

Recruiting talented bilingual American Indians proved to be the answer to the search for an unbreakable code. In World War I, British and American officers found that Cherokee troops in the 30th American Division could transmit indecipherable messages in real time, simply by using their native language. Choctaw Indians, trained by a U.S. Army commander to use their language in code, helped the American Expeditionary Forces win key battles in eastern France. The Choctaw communications had thwarted the efforts of German eavesdroppers to decipher the American messages.

In December 1941, the Japanese strike on Pearl Harbor brought the United States into the World War II. Within a few months, Philip Johnston, a civil engineer from California, proposed the use of Navajo code talkers in the Pacific Theater. The son of a missionary, Johnston had grown up speaking Navajo and recognized the language's structural complexity. He developed tests that proved trained Navajo code talkers could "encode, transmit, and decode a three-line English message in twenty seconds, versus the thirty minutes required by machines at that time," according to the once-classified results.[27]

At first, twenty-nine Navajo worked together to develop the shorthand vocabulary used by the code talkers in battle. The marines eventually recruited more than four hundred Navajo men, most of whom served in the Pacific.

American Indian participation in the military during the war outnumbered their percentage of the American population. Among them, perhaps the most famous are the Navajo code talkers, whose linguistic traditions and personal bravery contributed to the Allied victory in the Pacific.

One of the foods that would welcome the warriors back was Navajo fry bread. According to their oral tradition, the Navajo people created fry bread in 1864, using the flour, sugar, salt, and lard issued to them by the U.S. government. Forced to walk three hundred miles from their land in Arizona to New Mexico, many of them died along the way. In addition, the new land

In February 1945, on the strategic island of Iwo Jima, Navajo soldiers helped turn a bloody disaster into victory. Major Howard Connor commended the code talkers for their skill, speed, and accuracy throughout the battle, commenting, "Were it not for the Navaho, the Marines would never have taken Iwo Jima." Because the military classified the code as top secret, the government did not recognize the contributions of the code talkers until their work was declassified in 1968. Since then, Presidents Reagan, Bush Senior, and Clinton have honored these American Indian Marines at the White House.[28]

could not support the food crops on which the tribe's health depended, like beans, corn, and squash. To survive, the Indians used the flour, sugar, salt and lard to make a flat, fried bread they could eat. It provided little in the way of nutrition, but it kept their people from starving to death and allowed their traditions and culture to endure.

Visit a cemetery on an American Indian reservation, and you'll be impressed by the number of graves adorned with veterans' stars. The warrior tradition continues as a proud spirit among tribal groups.

Today, fry bread is served by Navajo and other tribal nations at home, at powwows, by vendors, and at restaurants throughout the Southwest. Nutritionists blame the lard used in frying the thick, flat bread for contributing to high rates of heart disease and diabetes among American Indians. Still, it is a delicious, handmade food that resonates with a long, vivid tradition. Fry bread welcomes every Navajo warrior home from his or her service in the military and from the white man's world.[29]

Navajo code talkers the United States employed during WWII to transmit and receive coded messages in their native language, which the Japanese could not decipher.

❧ FROZEN FOOD ❧

The concept of freezing food to preserve freshness and flavor was not born in the modern age. For millennia, Inuits flash-froze fish in the sub-zero temperatures of the Arctic winters. But an American businessman named Clarence Birdseye developed the process and distribution system that permanently changed the way people preserved food. Birdseye brought a new style of convenient, healthy foods to the American marketplace.

Birdseye conducted wildlife surveys as a naturalist working in Labrador, Canada, in the early 1900s. There he observed First Nation people allowing the bitter cold winter air to rapidly freeze the fish they caught. When defrosted for consumption, the fish showed very little deterioration.

In 1919, Birdseye founded the Birdseye Seafood Corporation, using a new product called "cellophane" to keep the food package airtight and thus protected from bacteria. His company grew organically with the growth of home refrigeration. In fact, seeing more opportunity in freezer distribution, Birdseye sold his company in 1929 to a new conglomerate named General Foods. The marketers at General Foods divided the last name "Birdseye" into two separate words—"Bird's Eye"—and pictured a little bird on the package.

Meanwhile, Clarence Birdseye began leasing deep freezers to local merchants. Flash-frozen food came on the market during the Depression. Most consumers could not afford it except as a special convenience, and the food markets could not afford the necessary display and storage units for frozen food. Birdseye made it possible for retailers to lease deep freeze equipment at a reasonable rate. Still, sales for frozen food limped along.

The real turning point came in 1944, during the last year of the war. Birdseye recognized that, for frozen food to be available everywhere, there had to be a distribution system that could keep the food below freezing. He began leasing insulated railroad cars, forging a national distribution network for frozen food.[30]

The timing couldn't have been better. With the return of the troops from the war, and a renewed optimism welling up in America, the marriage rate skyrocketed. Home sales and new construction hit record highs. The baby boom was launched. And young modern families were delighted to stock their shiny new freezers with frozen food. It was the start of a new era in food.

By the way, next time you wander down the frozen food aisle, look for the stylized image of a bird's eye on the front of a package of Bird's Eye frozen vegetables. You'll see Clarence's legacy staring back at you. Their frozen baby artichokes are really, really good. Try them sautéed with bits of smoked ham or bacon.

chapter 9

COCA-COLA, ICEBERG LETTUCE, AND FAST FOOD

The Postwar, Cold War Era

fter World War II, Americans faced an unpredictable new world. Life seemed so much better now that the Depression was past and the war was over, but worries continued. The fear of nuclear attack and the anti-Communist witch hunts marred the 1950s and most of the 1960s, with the Cold War igniting into the Korean War and later the Vietnam War. Strong leadership in the area of civil rights made progress against centuries of racial injustice, but it was a slow, hard, struggle.

It's easy to look back through a nostalgic viewfinder and see the entire era as an episode of the TV series *Happy Days*. With the GIs home, marriage rates and, inevitably, birth rates skyrocketed. The baby boomers made their first appearance, evolving into a new generation that would dominate social and technological change as never before. Their likes and dislikes eventually shaped fifty years of music, art, fashion, and yes, food.

BITE 81

Jell-O

Jell-O hit its peak of popularity in the Cold War era of the 1950s and '60s, but it was invented back in the Gilded Age. A man named Pearle Wait from LeRoy, New York, invented a type of instant, fruit-flavored gelatin that his wife, May, christened "Jell-O." It was 1897, and cute product names that ended in an *O* were the darling of advertising agencies.

A century earlier, the very idea of an easy, quick way to make gelatin at home would have surprised even the most experienced chef. Knowledgeable housewives simmered animal bones and cartilage for hours until the cartilage disintegrated. The flavorful stock would gel once it cooled. Known as aspic by the British, it was much admired by the French who called it *gelée*. *Oeufs en gelée*, or eggs in aspic, was a highly esteemed dish in Napoleonic France, and a nice bourgeois bistro in Paris will offer this dish to its clientele today.

In the eighteenth and early nineteenth centuries, elegant societies found transparent fruit gelatins a perfect dessert. Small V-shaped glasses, filled with brightly colored clear jellies, glowed like stained glass on étagères placed on the dining table. The jelly glasses reflected and amplified the candlelight, enhancing the entire dining experience.

Most people, however, did not have access to that kind of recherché treat until Charles B. Knox developed unflavored gelatin in 1894. At first, housewives and cooks bought it by the sheet from Knox's door-to-door salesmen,

---

Starting around 1600, the finest dessert gelatins, jellied candies, and quivering blancmanges got their jiggle from a very expensive product called "isinglass." A form of collagen, isinglass is obtained from the dried swim bladders of fish. One of the most remarkable qualities of isinglass is its neutral flavor, allowing this fish product to provide the basis of, for example, a delicate strawberry gelatin served at the White House by Thomas Jefferson.

who demonstrated how to make gelled desserts. Rose Knox, Charles's wife and business partner, published *Dainty Desserts*, a cookbook of recipes using Knox gelatin, and sales jumped.

Meanwhile, Pearle and May Wait experimented with their own version of gelatin by adding sugar and fruit flavors to their powdered gelatin. But they had little experience in marketing or sales and thus had trouble getting the product to take off. Discouraged, the Waits sold their patented Jell-O powder to a neighbor, Orator Francis Woodward, for $450 in 1899.

The Jell-O business continued sluggishly until Woodward launched an advertising campaign a few years later that gave sales lasting momentum. He called Jell-O "America's most famous dessert," and the tagline eventually became reality. Posters and magazine ads proclaimed the gelatin "America's Most Famous" and the product began to fly off the shelves. They printed and distributed fifteen million recipe booklets, some of them illustrated by the artist and illustrator Norman Rockwell, inspiring home cooks to give Jell-O a try.[1]

In 1924, Woodward renamed his food business the Jell-O Company and merged with Postum Cereal, ultimately forming the megacorporation

An early advertisement for Jell-O. The irony (or brilliance) of this is that Jell-O had only just begun to become popular, but they were already calling it "America's most famous dessert."

General Foods. With millions of advertising dollars in their coffers, General Foods embraced a startling new medium called radio to promote Jell-O. Jack Benny, one of the most popular radio hosts and entertainers of the day, nightly spelled out "J-E-L-L-O" over the airwaves during his show. People loved Jack Benny, and they grew to love Jell-O too. Even Eleanor Roosevelt enjoyed the dessert, and her wretched cook, Mrs. Nesbitt, served molded lime Jell-O with little marshmallows as a salad at the White House in the 1930s. Food at 1600 Pennsylvania Avenue had hit an all-time low.[2]

In the 1950s, as the postwar demographic eruption known as the baby boom was in full swing, mothers and grandmothers delighted their families with Jell-O. Long buffet tables at family reunions, church suppers, and Fourth of July celebrations were guaranteed to feature several Jell-O desserts and salads during the '50s and '60s. When kids ran around and bumped into the table, a good percentage of the dishes jiggled.

Today, Jell-O as a dessert still has many fans, although cupcakes, chocolate cakes, ice cream, and chocolate chip cookies challenge it in online popularity contests. It appears with regularity at cafeterias and, mixed partly with vodka instead of all water, at college drinking parties. Little kids still giggle when they have a spoon of Jell-O, because it's simply fun to eat.

BITE 82

☙ ICEBERG LETTUCE ❧

Today, people eat lettuce throughout the world and throughout their day. It may be in a breakfast sandwich, served as an appetizer, and featured as the main course for lunch or dinner. In postwar America, the lettuce most people ate was iceberg.

Since the days of prehistoric foragers, lettuce has provided nutrients and flavor to human meals. The ancient Egyptians cultivated it by 2500 BC, and from the land of the Nile it spread to Greece and, later, Rome. There it earned its Latin name, *Latuca*, from its milky juice. Lettuce then thrived in the cool kitchen gardens of medieval Europe. By the 1700s, amateur botanists had developed several cultivars, some of which are still available for us to enjoy today.

Crisp iceberg head lettuce ranks as the most popular type in America. Unlike other forms of lettuce that grow in green or reddish leaves, it is a crisphead, a pale, round ball of tightly wrapped lettuce leaves. The famous

American nurseryman, W. Atlee Burpee & Co., developed iceberg lettuce in 1894 and named it for its icy pale coloration. Plus the lettuce required a lot of ice to keep it cool during shipment. Otherwise, the crisphead did not travel well.[3] Postwar technology would change all that. After World War II, iceberg lettuce became a symbol of industrialized agriculture. The Whirlpool Corporation developed a form of controlled air conditioning and cold packing that extended the shelf life of this lettuce after harvest. This meant that iceberg from California could be shipped everywhere in the continental United States without spoiling. Now everyone could eat fresh lettuce at any time of the year. By the 1960s, 95 percent of all lettuce consumed in America was iceberg.

Huge lettuce farms thrived in places like Imperial Valley, California. The managers hired Mexican immigrants to tend and harvest lettuce, along with other crops. Following in the footsteps of earlier immigrants, these laborers worked the hardest jobs, occupying the lowest rung on the ladder for agricultural workers—migrant stoop labor. Many of them were undocumented and feared deportation if they complained about inhumane working conditions.

The lettuce workers in California eventually joined the United Farm Workers—a nonviolent labor organization led by Cesar Chavez, Dolores

Field workers harvesting lettuce in 1968.

Huerta, and Gilbert Padilla—that was born during the protests against inhumane working conditions in the California vineyards in the 1960s. In 1965, the UFW organized a boycott of California grapes, raising awareness of the farm workers' struggle.

Seven years later, the UFW encouraged a boycott of iceberg lettuce, which contributed to the success of their campaign. Ultimately, through their negotiations, they won the right of protection from dangerous pesticides sprayed on the plants, access to clean drinking water, and access to working toilets near their workplace. The humble crisphead of iceberg lettuce became a symbol of farm workers' rights.[4]

Today Americans enjoy many kinds of lettuce, but iceberg continues as the top seller with 60 percent of the market. That said, no longer an icon of new technology or a symbol of labor improvement, iceberg has lost a bit of its celebrity status. People enjoy its crispiness in sandwiches and tacos, as a Vietnamese summer-roll wrap, and in the classic wedge salad. It may be a little bland, lacking the flavor and nutrients of leaf lettuce, but it still adds a satisfying crunch to lunch.

BITE 83
❧ COCA-COLA ☙

The story of this addictively sweet soft drink begins with a Civil War veteran looking for an antidote for opiate addiction.

In 1886, Dr. John S. Pemberton created a distinctively flavored syrup that included flavorings from cola nuts (which added caffeine) and coca leaves (which are the source of cocaine, although this flavoring was removed in 1903). A pharmacist in Atlanta, Georgia, Dr. Pemberton was searching for a new, healthy soft drink that could be sold at drugstore soda fountains. He was also searching for an antidote for a morphine and opium addiction that had haunted him since being wounded during the Civil War.

The first people who sampled the syrup mixed with carbonated water immediately liked the taste. Pemberton's bookkeeper and business partner, Frank M. Robinson, named the soft drink "Coca-Cola," pooling together the names of the two major flavorings. He also invented the brand's trademarked script that the company still uses today.[5]

Two years later, Pemberton sold a majority interest in his soft drink to Asa Candler, a local businessman, who expanded the distribution of the syrup to soda fountains outside of Atlanta. (The inventor died soon after the sale, apparently from an opium overdose. Clearly his invention didn't help him as he'd hoped.) In 1899, Candler sold the bottling rights to a small group of innovative businessmen from Tennessee, who developed what became a worldwide bottling system. So many competitors were creating ersatz versions of Coca-Cola, however, that in 1916 the official bottlers designed a standardized contoured bottle, which effectively set Coke apart from the imitators.

Coca-Cola was not the first nor would it be the last carbonated soft drink invented in America. The late 1800s produced a wave of sparkling beverages, some of which still exist, each with their own "health" claim. Hires Root Beer, from Philadelphia, promised to purify your blood, while Dr. Pepper, named after a pharmacist in Texas, supposedly aided with digestion. Coca-Cola, according to Pemberton, helped fight headaches, exhaustion, impotence, and addiction to opium and morphine.

One of his first advertisements positioned it as a "temperance drink" and an "intellectual beverage." In fact, Coke is a great example of those patent medicines that proliferated in the late nineteenth century, making outrageous claims about their efficacy and healthfulness. Once government regulation required some semblance of truth in advertising and Congress began to tax medicine, the companies quickly dropped their boasting.[6]

Coca-Cola began its military career with the 1898 Spanish-American War. By 1900, men and women enjoyed a cocktail of rum, Coke, and a twist of lime called a Cuba Libre and toasted the success of American interventionism. Bottles of the soda went to Europe during World War I to revive the spirits of the American troops. During Prohibition, sales increased, since alcohol was banned until 1933 and carbonated beverages seemed more fun than water. A bottle cost a nickel in the Depression, a small extravagance that many could afford.[7]

Coke executives convinced the U.S. Congress that the soft drink was a wartime necessity early on in World War II. The company built makeshift bottling plants behind the front lines in Europe and the Pacific war zones and made sure Cokes were available at home bases. Not only were American soldiers encouraged to embrace the beverage, but the local populations in the war zones were encouraged to consume it as well. The wartime bottling plants became civilian bottling plants when peace finally returned.[8]

By the 1950s Coke was almost everywhere except for the Soviet Union and China, the two most powerful Communist countries and our enemies during the Cold War. Later, Coke would be banned in Cuba too. During the Korean War, American GIs had plentiful access when they were on base, as Coke installed vending machines at each camp post exchange, or PX. Today, North Korea is one of the last countries in the world where Coke is not sold legally.

Coca-Cola's effective use of advertising is one secret to its striking popularity. Like Jell-O, it embraced radio and TV early on as ideal platforms for mass marketing. The corporation also hired popular musicians and pop stars to appeal to a wide variety of Americans. Crossing racial barriers, they asked Ray Charles to sing their jingle. Charles was joined by a long list of pop artists in the 1960s singing commercials for Coke. The marketers realized that, even though a few adults still shook their heads over this phenomenon called rock and roll, music could help them reach out to the younger generation. Connecting with the youth market would be fundamental to Coke's success for the next fifty years.

COKE: A TRULY MULTIPURPOSE FOOD

By the way, Coca-Cola isn't only for drinking. People use it to glaze ham or marinate chicken wings, or as a dessert ingredient. Cherry Coke seems an especially popular addition to chocolate cake and brownie mixes. In the postwar era, a famous rock singer used it in what he called a salad. Elvis Presley (1935–1977) is almost as much of an American icon as Coke, and he certainly helped increase its fame. Early in his career, he told the press that one of his favorite foods was Coca-Cola Salad, a molded gelatin made of black-cherry Jell-O, Coke, crushed pineapple, pecans, and mini-marshmallows.[9]

BITE 84

★ PIZZA ★

Like Coca-Cola, pizza was not a new food in 1945, but it rapidly rose in popularity after World War II. It is practically the staff of life in the United States, even though it was invented in Naples, Italy. How did this come about?

The word "pizza" first appeared in print several hundred years ago in Italy. Fast-forward to the 1800s and a Neapolitan cook named Raffaele Esposito serving pizza to Italian King Umberto I and Queen Margherita during their visit to Naples in 1889. The classic pie with tomato sauce, mozzarella cheese, and fresh basil is named Margarita (in Italy, Margherita) because, according to legend, that was the queen's favorite among several choices of pizza.[10]

Southern Italians brought recipes for pizza with them to the United States, where the first pizzeria

In the 1950s, Hollywood's cool Rat Pack, which claimed Frank Sinatra, Sammy Davis Jr., and Tony Bennett as members, ate pizza. Another Rat Pack member, Dean Martin, sang lyrics that mentioned pizza in his popular song "That's Amore."

opened on Spring Street in lower Manhattan in 1905. Hungry young people found "pizza parlors" in Italian neighborhoods, but the food did not become a national phenomenon until World War II.

It was the soldiers returning from their experience in Italy that prompted the growth of pizza in America. Veterans who had fought Mussolini's troops put enmities aside as they fell in love with the simple deliciousness of an excellent slice of pizza. Pizzerias started to sprout up everywhere, even in small towns. Pizza became the easy food that cool people, including teenagers, ate. It was *in*.

Pizza was not the first fast food in America by any means. As we've seen, oysters win that title by a long shot. And home delivery on a small scale has also been available at least since the days of ancient Rome. But the widespread appearance of this once-foreign food in the postwar years was remarkable. Now it's available to order online, and city dwellers have it delivered to their houses at record speeds.

The success of Pizza Hut, established in 1954, and its rapid growth through franchising, is one factor that brought about the vast popularity of pizza. Domino's pizza delivery service is another reason why the Neapolitan pie is so successful. Domino's has ten thousand stores, corporate and franchised, in seventy countries. Small independent pizzerias operate in every city, and some of them serve outstanding homemade pizza. Walk down bustling Third Avenue in midtown Manhattan, and you'll spot a tiny, closet-sized pizzeria on almost every block.

> Today, Americans on average consume 350 slices of pizza every second.[11]

Pizzeria Uno in Chicago sold the first deep-dish Chicago-style pizza, to the endless delight of the inhabitants of the Windy City. And the local pizza in San Francisco is some of the best in the country. The best part about pizza is its easy adaptability to local flavors. Try pizza with chipotle in Santa Fe, New Mexico, or Barak Obama's favorite, pizza with ham and pineapple, on Maui. People are always creating new forms of pizza, making it part of their own culinary tradition.

BITE 85
TV Dinners

In many ways, TV dinners convey the modernist zeitgeist of the early post-war years as evocatively as any artifact or archive. Start with the fact that they were frozen and conveniently stored in that modern appliance, the freezer. Meanwhile, air travel for civilians was taking off. Advertisements for airlines featured chic and perky stewardesses with coy hats and bright red lips. On board, they served passengers free meals in divided, silvery trays that seemed so neat and "modern." These airplane meals inspired individual dinners frozen on trays.

TV dinners have an interesting origin. Those divided trays caught the eye of Gerry Thomas, a salesman at C. A. Swanson and Sons. In his very entertaining account, "Tray Bon!" author Owen Edwards reports that, in 1953, someone at Swanson "colossally miscalculated" the number of turkeys their customers would order for Thanksgiving. The company was stuck with 260 tons of frozen turkey, sitting in railroad cars. Inspired by the pre-prepared airline meal, Thomas ordered five thousand divided aluminum trays and had the sections filled with turkey with gravy and cornbread stuffing, sweet potatoes with a pat of butter, and green peas.[12]

Swanson had a hit on its hands. The following year, 1954, the company sold ten million turkey dinners. Soon new entrées like Salisbury steak and (soggy) fried chicken joined the frozen Thanksgiving favorite. The three separate sections of the meal were joined by a little square for cranberry sauce or even a serving of apple brown betty, depending on your choice. The best part for consumers? There was no prepping! No cleanup! Just take the package out of your swell new freezer and pop the tray into a preheated oven set for 350 degrees.

The streamlined convenience was a long, revolutionary road away from the labor-intensive meals with which most adults had grown up, and their baby boomer kids mostly just thought TV dinners were "neat." Home cooks everywhere breathed a sigh of relief at this miraculous new culinary invention.

Even the name evoked newness. TV dinners naturally were associated with that remarkable cutting-edge entertainment source, the television. In 1946, only .1 percent of households had a TV. By 1952, that number had grown astronomically to 53 percent, or 24.3 million households. The large, bulky boxes with small screens were more than a craze. For many American families, they had become the centerpiece of the home. Like the frozen dinners in disposable, divided trays, TVs represented the modern, the new, the convenient, and even the cozy.[13]

By naming their product "TV Dinner," Swanson created the perfect brand for the 1950s and 1960s. All over suburbia, in small towns and big cities, mothers were relieved, babysitters happy, and children delighted by every aspect of food on a tray. It even looked nutritious. Each tray section had its own member of a different food group. Today, you'll still find an overwhelming array of TV dinners in the freezer section of the supermarket, some claiming to be healthy and some delicious. Though no longer a cutting-edge culinary convenience, these easy, frozen meals are still just as popular as they were half a century ago.

TANG

A few years after Swanson invented the TV dinner, General Foods developed a bright orange powder called Tang as a breakfast drink for travelers. You just mixed the orange stuff into cold water and, bingo, a tangerine citrus drink. Sales were pretty dismal, however, until NASA decided Tang was just the thing to send into space with astronaut John Glenn. In 1962, Glenn was a true American hero, handsome and brave, helping our country to win the space race. (He was even better than a cowboy to many little kids—and his image was on a lunch box!) Tang orbited the earth with Glenn and later traveled on the Gemini space flights as well. Naturally, after Glenn took off into space with Tang, sales of Tang took off like a rocket.[14]

But not everyone loved the tangy powdered drink. At least one of the three astronauts on the historic Apollo 11 flight to the moon in July 1969 was not impressed by the orange powder NASA had sent along with them. When asked about the food during his mission in a 2013 interview, Buzz Aldrin observed, "Tang sucks."[15]

BITE 86

⚘ CUBAN ⚘ AMERICAN FOOD

In 1962, America looked markedly different than it had ten years earlier. Little girls now played with a provocative new doll named Barbie, who began to supplant traditional baby dolls. Hula-Hoops and thick milk shakes had won the hearts of young adolescents, while older teens sang along with a new band called the Beach Boys and twisted to 45s by Chubby Checker.

More significantly, the U.S. Supreme Court upheld racial integration on buses, on trains, and in schools during the 1950s. Injustice could no longer be swept under the rug quite so easily, partly because journalists started covering civil rights stories. The space race against the Soviets was in full swing. Astronauts were a new type of hero. There was a young new president in the White House named John F. Kennedy. And the possibility of a nuclear war had never been more real, thanks to an event called the Cuban Missile Crisis.

Earlier, in 1956, the Cuban American population hovered around 125,000, largely based in Miami and in the cigar-manufacturing district of Tampa known as Ybor City. Cuba itself was the fiefdom of a despot named Fulgencio Batista, who had once been popularly elected. After losing his second-term election, he returned to power by staging a military coup. The country reeked of corruption, thanks in part to the Mafia and American corporate policies. Jack Kennedy, still just a senator from Massachusetts, described Cuba under Batista as a "complete police state."

Fed up with oppression, a band of socialist revolutionaries, led by Fidel Castro and Che Guevara, began a three-year guerrilla war against the Batista government that ended in victory for the radicals in January 1959. One month later, Castro became the new prime minster at the age of thirty-two. The new government established free primary schools and medical care for all Cubans. It also nationalized factories and farmland.

When the United States froze Cuba's assets and instituted a strict embargo, the Soviet Union stepped in as Castro's most supportive ally. The Soviets stationed nuclear weapons in Cuba in October 1962, which triggered that terrifying moment in radioactive brinksmanship known as the Cuban Missile Crisis. This thirteen-day confrontation was the closest the world came to a nuclear conflict during the Cold War. Intense negotiations between President Kennedy and Soviet Prime Minister Nikita Khrushchev ultimately ended in a tense peace between the two world powers, punctuated by proxy wars in developing nations.

Cubans from all walks of life—from surgeons and engineers to cooks and taxi drivers—fled Cuba after the revolution. The number of Cubans in the United States tripled to more than three hundred thousand. They brought new styles, new music, and new food to American culture.

One of the most famous foods the new immigrants brought with them is the well-known Cuban sandwich, or *media noche* (middle of the night), consisting of sliced pork, ham, and cheese, topped with pickles and mustard on egg bread. Cuban cooks marinate poultry and red meats in citrus juices, then slow roast them over a low flame, so that the meat almost falls off the bone. The colorfully named "Moors and Christians"—black beans and white rice—is also known as *congri* and frequently finds a place on Cuban tables. The food is not too spicy, but very satisfying and filling.

OFELIA BRAGA'S
PICADILLO CRIOLLO

Ofelia Braga arrived in Miami in 1960. She learned to cook growing up in a farm family that had become desperately poor under the Batista government. When the Cuban Revolution came, she hoped her life would change, but Castro's new policies did not offer her the opportunities she was looking for as a young woman. She settled in the Little Havana neighborhood of Miami, and with her husband, raised a large family. Today, Ofelia still cooks Cuban food and runs her kitchen like a warm-hearted commandant. Her daughter, Marlene, is a TV producer and a dear friend of mine.

Marlene says this is a traditional dish in her family. It includes sofrito, *the onion, green pepper, and garlic mixture that adds basic flavor to many Cuban dishes. Sometimes* sofrito *includes oregano or cumin, but Mrs. Braga's recipe leaves those seasonings out. Typically, the picadillo is served with a big pile of white rice. Rice is another basic component of a Cuban meal. Marlene recommends a simple salad of avocado and iceberg lettuce as a nice accompaniment.*

If it were my recipe, I'd brown the beef before I added the other ingredients and I'd chop up the olives. Maybe I'd add cumin and oregano. But it is not mine. It's the resilient Mrs. Braga's, so don't mess with it!

INGREDIENTS

- ► 2 pounds lean ground beef
- ► 1 small yellow onion, peeled and chopped
- ► 4 cloves garlic, peeled and minced
- ► ¼ seeded green bell pepper, chopped
- ► ½ cup tomato puree

- ▶ ½ cup dry vermouth or white wine
- ▶ ½ cup pitted green olives, preferably with the pimento center
- ▶ ½ cup raisins
- ▶ ¼ cup ketchup
- ▶ Salt and freshly ground black pepper to taste

DIRECTIONS

Mix the ground beef, onion, garlic, bell pepper, tomato puree, vermouth, olives, raisins, ketchup, salt, and pepper together in a large pot over low heat. Cover and simmer for 30 to 45 minutes. Stir frequently.

Serve with a big pot of white rice and a platter of sliced plantains you fried in hot oil while the picadillo was simmering gently.

Serves 6 as a main dish.

BITE 87
JACK KENNEDY'S FISH CHOWDER

The Cuban Missile Crisis was one of the defining moments of the Kennedy administration. John F. Kennedy remains an iconic president in American history, not just because of his good looks, youthful vigor, or apparently heroic military service. The early 1960s were a time of great promise, and for many, JFK personified that moment of American energy and idealism. Historians will debate his actions during the Cuban Missile Crisis, his obsessive womanizing, his chronic use of painkillers, and the effectiveness of his domestic policy. But there is no denying his inspirational leadership during the three short years of his presidency.

His elegant wife, Jacqueline Bouvier Kennedy, was an equally iconic figure. A natural beauty and an utterly chic First Lady, Jackie was an asset in foreign relations and an ardent preservationist. She championed the arts and helped restore the White House's historic interiors.

She also restored the reputation of the White House kitchen, which had never quite recovered from the tyranny of Eleanor Roosevelt's grim Mrs. Nesbitt. By hiring French chef René Verdon, Jackie Kennedy returned the White House to the gourmet traditions of Thomas Jefferson and Dolley Madison. Mrs. Kennedy never claimed to be an in-the-kitchen cook, but she was a formidable hostess of enormous talent, presiding over state dinners and elegant private dinners with her husband. The elaborate menus featured dishes with names like *Poulet à l'estragon* and strawberries stroganoff, which sounded quite sophisticated to the American ear and palate.[16]

In contrast, President Kennedy was not personally interested in food and frequently had to be reminded to eat. He preferred a bowl of soup and a sandwich at lunch. His favorite soup was chowder, made with clams or fish, appropriately enough for a man who called Boston and Cape Cod home.

Fish chowder is a traditional old French soup, using the catch of the day as the main ingredient. Onions, carrots, herbs, and stale bread crumbs were combined with fish stock and other seafood in a large pot, known as a *Chaudière* (hot pot). By the 1800s, potatoes became a common addition, frequently replacing the bread crumbs. In America, New England cooks finish the chowder with milk, while in New York, diced tomatoes are added. As soon as Kennedy opened his mouth, you could tell he was not a native New Yorker. His tastes reflected that as well: his favorite chowder was always based on New England traditions.

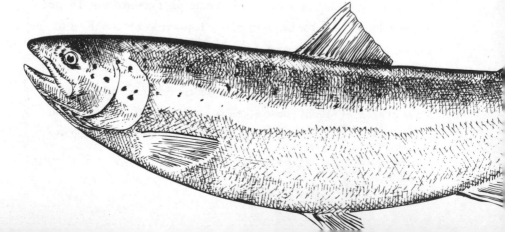

NEW ENGLAND FISH CHOWDER

This recipe for fish chowder is similar to the one JFK enjoyed for lunch at the White House. He also liked the ever-popular clam chowder, but the timing on that can be tricky. If you've never cooked fish chowder, give this recipe a try.

INGREDIENTS

- 1½ pounds haddock fillets
- 1 pound other fish fillet, such as salmon, cod, skate, or a mixture
- 2 cups water
- 4 slices smoked bacon, diced
- 2 medium onions, diced
- 1 tablespoon olive oil, if needed
- 4 large potatoes, diced
- ½ cup diced celery
- ½ cup diced and peeled carrots
- 1 bay leaf
- 1 teaspoon dried thyme
- 1 teaspoon salt
- Freshly ground pepper to taste
- 1 cup fish stock or clam juice
- 4 cups 2 percent milk or whole milk
- 2 tablespoons butter

Simmer the fish fillets in water for 6 minutes in a covered pan. Drain, reserving broth. Cook the bacon in a large pot over low heat until crispy and the fat is rendered. Remove the bacon, leaving the fat in the pot, and drain on paper towel. Set aside.

Sauté the onions in bacon fat until golden. Add olive oil if needed. Do not brown. Stir in the potatoes, celery, carrot, bay leaf, thyme, salt, and pepper. Add the reserved fish broth plus fish stock or clam juice. Bring to a boil and then immediately lower the heat to simmer. Cover pot and let the chowder simmer about 20 minutes or until potatoes are tender.

Cut the fillets into small pieces. Add to soup. Simmer 5 more minutes. Add the milk and butter, and heat thoroughly over low heat. Do not bring to a boil. Adjust seasonings.

Serve chowder in warm bowls, with crispy bacon crumbles on top of each bowl. Enjoy while playing a game of Risk.

Serves 6 to 8.

BITE 88
⚘ CRÈME CARAMEL ⚘

About the same time that Jackie Kennedy hired a French chef for the White House, Julia Child first introduced Americans to French cooking. All over the country, aspiring gourmets traded plain beef stew for Boeuf Bourguinon and canned condensed soup for a homemade potage. They were delighted with the results. Julia Child's timing was exactly right. A strong cadre of Americans was ready to take more than one step up the culinary ladder from SPAM and Jell-O.

Julia Child's *Mastering the Art of French Cooking* appeared in bookstores in 1961 and was warmly embraced by home cooks all over America. The clear, friendly writing and delicious results inspired the competitive cooking instinct among some housewives who had few outlets for their ambitions in the 1960s. (My own mother produced a delectable vichyssoise for her dinner parties, having whirled cooked potatoes and leeks into a silky puree in her one-speed Waring blender.)

Ten years later, *The French Chef* TV show debuted, starring the delightful Mrs. Child herself. She bustled about her own kitchen, demonstrating cooking techniques with innovative overhead camera angles so that the television audience could actually see what the chef was doing. She spoke to the camera as if it were her friend, which appealed to Americans everywhere, and matter-of-factly shared her expertise with confidence and skill.

This long-running public television series inspired people to expand their culinary repertoire and their family menus and, most of all, their palates. No longer resigned to the bland flavors of the processed foods the United States was known for, Americans began experimenting with new foreign dishes and tastes. The national palate was growing more sophisticated.

Professional cooks were inspired as well. Julia Child challenged American restaurants to up their game. With Mrs. Kennedy and Mrs. Child simultaneously championing gastronomic excellence, and with such different personal styles, it is not surprising that culinary standards improved. Julia Child's successful TV shows were a precursor to the cooking programs on popular cable networks like the Food Channel today. In 2009, the movie *Julie and Julia*, starring Amy Adams and Meryl Streep, offered a fabulous rendition of the doyenne of French cooking in America and brought the magic of Julia Child to a new generation.

MY LOVE AFFAIR WITH CRÈME CARAMEL

Instead of waxing on about how revolutionary and inspirational she was for American cuisine (which I'm sure most, if not all of you know), I want to share my own experience in discovering the magic of Julia Child's culinary expertise and gentle instruction.

Her recipe for crème caramel inspired my first independent foray in the kitchen. I taught myself how to make crème caramel when I was ten. I had been to France three times, traveling with my family in a Volkswagen bus on lengthy summer holidays. My father, a history professor, would spend weeks working at the Bibliothèque Nationale in Paris while my mother took us to museums or supervised our play in the public gardens.

Then we would load up the van for a holiday, driving throughout France and neighboring European countries. In 1964, we listened to Radio Luxembourg on the car radio and heard a new band called the Beatles. My fifteen-year-old sister fell in love with Paul McCartney. We all read Gerald Durrell. And I developed an inordinate fondness for crème caramel, the burnt sugar custard served at every corner bistro and the cheap but homey pensions where we invariably stayed.

So, back home, I opened up *Mastering the Art of French Cooking* and made the custard myself. I learned how to use a bain-marie and melt sugar in a heavy pan. I ended up with a large brown scar, a burn from the melted sugar, which I had spilled onto the inside of my arm. It hurt like crazy but I kept mum. I feared the recipe would be placed off limits for my feats of independent cooking if anyone knew about the burn. In retrospect, any recipe that requires melting sugar and pouring it should be supervised by an adult, but my mother left me happily alone, unaware of any dangers lurking behind crème caramel's sweet facade.

That first crème caramel came out perfectly, and I've been baking it (in a bain-marie, of course) ever since.

Fast-forward twelve years to when I was twenty-one and a senior in college. A handsome young musician, whom I had admired since grade school, returned to our town for Thanksgiving with his family. He dropped by my parents' house over the holiday, and I served him crème caramel. It was love at first bite. And so, dear reader, I married him.

❧ MCDONALD'S ❧

One of the peculiar things about the United States is that the interest in gourmet cooking grew simultaneously with the spread of fast food. You might expect them to be mutually exclusive, but our culinary culture is varied and complex, to say the least. Just as Julia Child popularized French food in the 1960s, around the same time the development of fast-food chains exploded nationally and internationally, starting with McDonald's.

In 1948, brothers Dick and Mac McDonald opened an octagonal drive-in restaurant in San Bernardino, about sixty miles east of Los Angeles. Impressed with Henry Ford's assembly-line approach to production in the early part of the twentieth century, the brothers broke preparation, cooking, serving, and cleaning into simple steps. They treated their all-male teenage employees like new army recruits, setting specific timing and sanitary standards to be achieved with each step.

Dick McDonald soon designed a red and white building with golden arches, aiming to make their drive-in—and its classic American foods of hamburgers, shakes, and fries—stand out along the road where other fast-food stands were also competing furiously for customers. It was a profitable business; the brothers expanded to ten franchises.

One day, a salesman who sold the McDonalds their milk-shake mixers dropped by. His name was Ray Kroc, and he wanted to meet the owners of this new company that ordered so much of his equipment. Impressed by what he saw, Kroc bought the right to sell McDonald's franchises in 1955. Six years later, after selling more than one hundred licenses to run a McDonald's restaurant, he bought the whole company for $2.7 million.

That was a huge price for a fast-food company in 1961, but Kroc had a huge vision. He didn't invent fast-food hamburgers with fries and a shake, or the golden arches. But he did catapult a brand into a universal symbol for easy, reliable, speedy, inexpensive food, designed to appeal to all ages.

Kroc was as obsessed with cleanliness and thrift as the McDonald brothers he'd just bought out. He personally picked up any litter he found outside his McDonald's restaurant in Des Plaines, Illinois. He required all employees to be neat, clean, and never chew gum. Young women need not apply as servers, because they might attract young men in leather jackets, unsavory types who smoked. And that would make it just another burger joint, not a family place that mothers and wives would enjoy.

This was important: Kroc realized that his main customers were young families with kids. The food was soft and palatable to children and cheap to buy; there were no long lines and no cleanup. McDonald's was a paradise for a beleaguered parent in search of a family dinner after a long day at work. One of Kroc's first advertising slogans was "Give Mom the Night Off!" which later evolved into "You Deserve a Break Today."[17] The brand boomed, eventually becoming a truly global company.

Today, you can find McDonald's all over the world, from America to Austria to Afghanistan. They often serve the same type of fast-food fare in every location, but occasionally you can find food that reflects more of the local flavors. For example, in India, you can find the McAloo Tikki Burger (a potato and chickpea patty with onions, tomatoes, and tomato mayonnaise), in France some of the places serve Camembert Premiere (fried French cheese nuggets), and in Canada you can sometimes find the classic Poutine (french fries and cheese curds doused in gravy) on the menu.[18]

More recently, McDonald's has attracted critics by the thousands. In their defense, the corporation has been responsive to some complaints, for example, by reducing pollution, improving its beef-buying practices, and even offering salads and other healthier options on its menu. Much of McDonald's menu would not be healthy to eat too frequently, or in the extra-large portions for which modern fast food is famous (or infamous). But the food continues to be very, very popular, even in countries where a new McDonald's inspires protests.

SLOW FOOD VERSUS FAST FOOD

Despite its continued popularity, the fast-food culture epitomized by McDonald's angers many American and international social critics and has inspired documentaries, vandalism, and worldwide social responses such as the Slow Food movement. Originating in Italy in 1986, Slow Food is now an international organization of 150,000 people who value consuming seasonal food that is locally procured. They value the time spent in the kitchen and around the dining table.[19]

When Ray Kroc died in 1984, McDonald's held 7,500 restaurants worldwide. Today, McDonalds serves fifty million customers a day.[20]

I am a dues-paying member of Slow Food, and I understand the health dangers in consuming too much fast food. However, there are times when a McDonald's run hits the spot. I remember driving with my big brother, Terry, toward Allentown, Pennsylvania, where a funny-looking restaurant had opened on a local boulevard. Terry had recently passed his driving test, and it was so exciting to sit up front next to him on the bench seat of my mom's wood-paneled station wagon. It was 1963, and I was nine. My big brother ordered us hamburgers, and there was no waiting for our burgers and hot french fries, thin, salty, and so yummy. Then Terry handed me a chocolate shake so thick it made the striped plastic straw stand up straight. I had found heaven, and its name was McDonald's.

BITE 90
❧ SOUTHERN ❧
FRIED CHICKEN

Although fast-food chains also serve deep fried chicken, homemade fried chicken has a lighter taste. It is a symbol of southern cooking traditions, and it was the favorite of our most famous civil rights leader, Dr. Martin Luther King Jr.

Reverend King was a young minister, newly arrived in Montgomery, Alabama, when the gentle activist Rosa Parks quietly refused to give up her seat for a white person on a city bus in 1955. Her arrest for violating the segregation laws initiated a massive legal case that would travel up to the U.S. Supreme Court. The local African American community responded with mass meetings of the new Montgomery Improvement Association, led by Reverend King, Ralph Abernathy, and other civil rights leaders. It also organized an effective bus boycott, which lasted for more than a year.

Georgia Gilmore began walking to her job at the National Lunch Company as part of her participation in the boycott. She was proud to be part of the struggle for equality and attended the mass meetings which she said she heard about on the "Negro radio." In an interview recorded in 1986, she remembered, with more laughter than outrage, that white people would drive by as the protesters walked to work and call them names. "A lot of times, some of the young whites would come along and they would say, 'Nigger, don't you know it's better to ride the bus than it is to walk?' and we would say, 'No, cracker. No, we rather walk.'"[21]

> *King called Georgia Gilmore's three-hundred-pound cook "Tiny," according to Mark Gilmore Jr., Georgia's son. But it was clear her food occupied a big place in Reverend King's heart.*

Gilmore's employer at the Lunch Company fired her for participating in the boycott, attending the meetings, and—worst of all—testifying in court against a particularly nasty and racist bus driver. A mother of six, she knew she had to find a new job fast, but her options were limited. Her work as a part-time midwife brought in little money. Her other talent was cooking, and her southern fried chicken was locally famous. Reverend King helped her establish an informal restaurant in her house, where she cooked and served traditional southern food much like she had made at the National Lunch Company.

King brought his colleagues and associates to eat at Georgia Gilmore's place. He liked that he could meet people there confidentially, without outsiders listening in on his conversations. In later years, he brought presidential candidate Robert Kennedy, President Lyndon Johnson, and even die-hard segregationist George Wallace. But what he really loved was the food. And there was no meal Reverend King liked better than Gilmore's southern fried chicken, collard greens, sweet potatoes, corn bread, watermelon, and pecan pie. That was his ideal spread, and that's what she made for him and his friends.[22]

Gilmore also started an organization called the Club from Nowhere because that anonymity protected the club's members, all of whom sold homemade cakes and pies to raise money for Reverend King's movement. As the sole officer of the club, she brought the cash—sometimes $100, sometimes $250—to the mass meetings, where it was sorely needed.

In November 1956, the U.S. Supreme Court ruled against the city of Montgomery in a landmark decision, finding segregation on public transportation unconstitutional. The successful result of the bus boycott gave Martin Luther King the platform to become a national civil rights leader. He recognized the importance of everyday participants like Georgia Gilmore by saying, "Our work could not happen without the ground crew."[23]

A plaque now stands in Montgomery, Alabama, recognizing the work and cultural legacy of the cook who ran the Club from Nowhere and dedicated her tremendous culinary talents to equality and justice. Plus, she made the best southern fried chicken in town.

SOUTHERN BUTTERMILK FRIED CHICKEN

....

Deep-fried food is another example of a southern kitchen tradition influenced by African roots. The Scots and English settlers in the British colonies did not use a lot of oil in their cooking. Enslaved Africans, however, had a long tradition of deep frying in palm oil.[24]

If you are in Montgomery, Alabama, I hear that you should eat at Martha's Place, run by a talented southern chef, Martha Hawkins, who is one of the heirs to Georgia Gilmore's gastronomic legacy. The menu on her paper place mats says that she serves southern fried chicken every day. But if you can't get there, here is a recipe that I think is pretty good.[25]

INGREDIENTS

- 3¾ cups buttermilk
- 2⅔ tablespoons salt
- 1 tablespoon sugar
- 3 pounds chicken pieces (breasts, legs, and thighs)
- 2 cups all-purpose flour
- 1 teaspoon freshly ground black pepper
- 1 teaspoon dried thyme
- 1 teaspoon ground sage
- 8 cups peanut oil, or enough to fill about ¾ inch of skillet

DIRECTIONS

Combine 3 cups buttermilk, 2 tablespoons salt, and sugar in a mixing bowl. Pour the marinade into a 2-gallon zipper-lock bag. Add the chicken pieces, seal bag, and shake to coat. Make sure all chicken is coated. Chill for at least 3 hours or overnight.

Remove the chicken from brine and drain, then wipe dry with paper towels. Throw out the buttermilk marinade. Stir the flour with the remaining ⅔ tablespoon salt, pepper, thyme, and sage in a large bowl. Put the remaining ¾ cup buttermilk in a shallow pan. Coat each chicken piece with the flour mixture, shake off excess, and dip into buttermilk. Then dip back into the flour mixture. Let the coated chicken pieces stand for 20 minutes or so if possible.

Heat the oil to 350°F in a 12- to 14-inch heavy skillet or a deep-fat fryer. Using tongs, place a few pieces of chicken into the fryer and fry for about 15 minutes, until the coating is golden brown and the meat is no longer pink. Turn once during the cooking. The oil should be kept around 325° to 350°F. Drain the pieces on paper towels. Keep pieces warm in a 275°F oven while you are cooking the other pieces.

Serves 6.

It's better to use smaller pieces from younger chickens. Cut the breasts in half so that they cook evenly. Note that the thighs take a little longer to cook than the breasts.

chapter 10

MICROWAVE POPCORN, MESCLUN GREENS, AND SALSA

1969–2000

inally we have reached the era that most readers will remember, and they will be able to bring some of their own memories to the story of American food. While you are living through a time of change, it's hard to identify how our society has transformed without taking a step back to gain perspective. So consider what you had on your plate twenty years ago. Many of the foods have stayed the same, but new additions, once startling, have become common, if not mundane. Now let's take a look at the American plate in the last third of the twentieth century.

BITE 91

MICROWAVE POPCORN
(*Zea mays everta*)

My first chapter begins with maize, so it seems only fitting that corn leads the last chapter as well. After all, as we've seen, corn is one of the most American of all foods. The subspecies of popcorn maize, *Zea mays everta*, is at least two thousand years old and so named because the kernel everts, or turns inside out, when exposed to heat.[1] Other species might pop open somewhat, but only *everta* produces the individual white, fluffy clouds. By the late twentieth century, not only did we pop corn, we microwaved it.

Originally, American Indians enjoyed eating popcorn and introduced it to Europeans, although the exact place and date when this occurred is not clear. We can clearly trace the increase in popcorn consumption, however, by following technological advances over the past 150 years. For example, a new technology in the 1830s—strong iron wire—was woven into a fireproof covered basket at the end of a long handle. This corn popper contained the flying kernels as they cooked and exploded, but you could still see the action.

Popping corn became a central social activity for young people and family groups as they gathered around the hearth or an open fire outdoors. (You can still buy variations of these poppers at outdoor stores.) The handle of the corn popper was passed from hand to hand, so that everyone participated in the process. This also allowed courting couples to sit right next to each other in front of the fire.

Besides eating the popcorn from a bowl, hot and fresh from the popper, women and children made popcorn balls with cooked molasses or maple syrup. They could also string the popped kernels together, often alternating them with cranberries where available, to make decorations for the newly popular Christmas trees during the Victorian period. By the time of Lincoln's election in 1860, people all over the United States ate popcorn.

In 1885, another new technology transformed popcorn consumption outside the home. The steam-powered popcorn maker, invented by Charles Cretors, brought the snack to outdoor events at stadiums and arenas. Vendors at baseball games, circuses, and fairs all offered paper sacks filled with hot popcorn. Cretors's invention was noisy and dirty, but so were a lot of the places where his machine was used.

An electrical engineer from Montana named Charles T. Manley invented the third technology that sent popcorn sales flying off the cob. In 1925, he invented the electrical corn popper, which had a clean engine suitable for indoor use. In the 1920s, people bought popcorn outside movie theaters and brought their snack inside, so Manley's invention meant that theaters could cook and sell their own popcorn inside their own venue.

When the Great Depression hit, many theaters cut their admission prices because people couldn't afford the tickets otherwise. Popcorn sales were what kept their doors open. Even in bad economic times, most audience members could afford a bag of popcorn. And when the war came, Manley sold his machines to military bases and USO centers, bringing a taste of home (or at least a taste of neighborhood movie theaters) to people in uniform.[2]

A fourth development, the self-expanding aluminum pan known as Jiffy Pop, grew alongside television during the 1950s and '60s. Children could shake the pan over the stove burner and watch the shiny dome expand as the kernels exploded inside. When the popping stopped, a babysitter or parent could cut open the top and pour the popcorn into a big bowl. Like a TV dinner, Jiffy Pop was modern in its convenience and its disposability. No washing pots and pans for Mom. Jiffy Pop was swell.

Kitchen microwave ovens arrived in the 1970s, taking popcorn to a whole new level of speed and convenience.[3]

Around the same time, a scientist named Percy Spencer demonstrated the culinary power of microwaves to his colleagues at Raytheon by using the energy to make popcorn. The puffed kernels flew around his laboratory during his trial presentation. It took twenty more years before microwave technology became reasonably affordable and compact enough to become a standard kitchen appliance. Once again, American popcorn consumption took a big leap forward.

Today, one billion pounds of popcorn are consumed every year, and about 65 percent of that is microwaved and eaten at home. Families in the 1800s poured liberal amounts of hot melted butter over their popcorn, but of course they led more active lives than people do today and burned off those calories easily. The real health hazard isn't popcorn or a small amount of butter, but a chemical additive called diacetyl, an artificial butter flavor that has been found to cause lung cancer in workers exposed to these fumes at the popcorn factories. "Popcorn lung cancer" (bronchiolitis obliterans) has also afflicted consumers who inhale the aromas of a freshly opened bag of their buttery-flavored snack.[4]

However, if you just buy the kernels straight, pop them yourself, and season them however you wish, popcorn can become a relatively healthy snack. The American Indians enjoyed it that way, and we can, too.

BITE 92
❧ WONDER BREAD ❧

Humans have baked bread for thousands of years, so it's no surprise that the size, color, shape, and taste of this food that has been called the staff of life has varied markedly throughout the ages. I wonder what the ancient Romans would have thought of Wonder Bread, though.

While its ingredients have varied regionally and by economic class, bread has played a central role in western Europe throughout recorded history. By the Renaissance, the color and texture of the bread regularly on your table indicated social position and wealth. Peasants and laborers in Europe consumed dark, coarse bread while the upper classes ate paler, finely textured bread.

The reason behind this distinction? Millers sold highly refined wheat flour, well-sifted, at a higher price, and only wealthy people could afford this on a regular basis. Thus so-called "refined people" consumed "refined" bread. Their social inferiors ate whole-meal bread with mixed local flours such as rye, wheat, barley, or oat, depending on the region.[5]

The first European settlers in the New World longed for the wheat bread they relished in Britain or Spain, but settled for maize johnnycakes or tortillas until their wheat crops thrived. By the 1800s, Americans enjoyed Indian corn bread but also ate wheat bread. Thrifty New England bakers stretched their wheat flour with cornmeal, added molasses as a sweetener, and created delicious anadama bread. The laboring poor—free, enslaved, and indentured—might eat corn pone exclusively. Once again, white bread indexed social rank. The whiter the bread you ate, the higher your status was.

Most families baked bread at home or bought it from small, local bakeries. Eventually, urban ethnic enclaves included bakeries that sold breads with similar aromas and appearances to those in the old country, although the flour was not always the same. All bread was artisanal: if not homemade, it was handmade. That is, until the 1890s, when factory-made bread became more available.

In contrast to today's perceptions about wholegrain, handmade bread, factory-made loaves held the reputation of being healthier and more sanitary for decades because they were made of that oh-so-desirable refined flour, with dough kneaded by machines. People wanted clean bread that wasn't tainted by human touch. Bit by bit, industrially produced bread edged out home or locally baked bread.

The new white bread represented not only convenience but the modern American way. The same reformers who urged the passage of the Pure Food and Drug Act campaigned against "squalid" immigrant bakeries with "unscientific" techniques. Leading publications such as the Atlanta Journal-Constitution *and* Scientific American *denounced handmade bread as a health hazard.*[6]

Part of this was because an increasing number of women began working outside the home and lacked the necessary time to make bread for their family. "Boughten bread" checked at least one thing off a working woman's endless to-do list.

In 1921, a new type of factory bread emerged, made from "batter, not dough!" by the Taggart Baking Company in Indianapolis. The batter meant that no human ever touched the bread, thank goodness. It was pure white, it was spongy, and the manufacturer called it "Wonder Bread." A few years later, Continental Bakery acquired Taggart and began distributing Wonder

Bread nationally. It became one of the first breads to be sold and packaged *presliced*. Women loved the convenience and the low price, and sales continued to climb.

Overall, the amount of bread consumed rose during the Great Depression, but its nutrient levels declined when factories began to bleach flour to get it super white. Soldiers drafted for World War II manifested vitamin deficiencies because they were no longer getting the nutrients needed, even though they ate more bread than ever before. Alarmed by the increase of nutritional diseases like beriberi and pellagra, the federal government encouraged the enrichment of white bread with vitamins and minerals similar to those found naturally in whole grain bread.

After World War II, the average American ate one and one half pounds of white, presliced bread on a weekly basis. Although not all of that was Wonder Bread, it was generally a batter-based spongy loaf.[7]

The purity of refined or white wheat flour was still so coveted that it seldom occurred to anyone to try going back to the old methods of creating nutritious, hearty bread from vitamin-packed darker grains. Wonder Bread adopted the recommendations for fortifying bread and turned them into a marketing platform. By the early 1960s, their advertising campaign claimed that "Wonder Bread builds strong bodies twelve ways" due to twelve added nutrients. As advertising executive Jef I. Richards observed, "Advertising is the 'wonder' in Wonder Bread."[8]

All over America, baby boomers' moms packed school lunches with sandwiches made with Wonder Bread. It was the normal bread to eat—normal people ate Wonder Bread in an era when social conformity was very important. The brand's popularity crossed racial, generational, and class divides.

By the 1970s, Wonder Bread, now owned by Hostess, had morphed into a cultural symbol of conservative America. Archie Bunker, the lead character in the hit TV comedy *All in the Family*, probably ate Wonder Bread while condemning minorities and women's libbers. Most real families ate it too—it was well priced, easy to find, and predictable. Meanwhile, college students, social reformers, and cultural critics began to turn up their collective nose at the tasteless white loaf. To them, it represented all that was wrong with American food—industrially processed, chemically treated, mass-produced, and soul-less.

Wonder Bread sales began to fall in the late twentieth century. Influenced by health reports and changing trends in gastronomy, an increasing number of people again sought out brown bread with whole grains. By 2010, for the first time in decades, more Americans ate whole wheat bread than white bread. Two years later, Hostess Bakery declared bankruptcy. A symbol of twentieth-century processed food, Wonder Bread was suddenly no longer available in the United States.

In 2013, Flowers Foods bought the rights to many of the Hostess products and began distributing Wonder Bread once again throughout the country. According to various food-focused message boards, a surprisingly large group of nostalgic eaters are enormously relieved. The true wonder is that so many Americans are now devoted to eating whole grain bread, reversing the eating habits of centuries.

BITE 93

GRANOLA

Made of whole grains with nuts and dried fruits, granola was born at the opposite end of the health-food spectrum from Wonder Bread. Rock music accompanied its modern rebirth at Woodstock, one of the great outdoor festivals of all time.

In 1968, the year before Woodstock, the whole country seemed to be exploding. Martin Luther King Jr.'s assassination in April lit an angry conflagration in 110 cities around the country. Louisville, Kansas City, Chicago, Newark, and Washington, DC, witnessed tremendous destruction and violence. The race riots left whole neighborhoods destroyed and changed old communities forever. The assassination of presidential candidate Bobby Kennedy stunned the nation a second time. Then the bloody Chicago riots during the Democratic National Convention followed in August.

Revolutionaries and reactionaries shouted their provocative slogans at one another. Generational conflicts grew bitter, especially between some members of the World War II generation and their long-haired children, who looked at each other across the dining table in hostility and bewilderment. The slim victory of Richard Nixon in the November presidential election only further alienated peace advocates who ranged from middle-class moms to counterculture hippies and Black Panthers, as the new president committed more troops to the escalating conflict in Southeast Asia.

At its start, 1969 didn't seem much better. But something happened that summer, an amazing moment of musical celebration that gave some hope. A somewhat chaotic and messy concert in Bethel, New York, provided a respite from all the fiery hatred. In August, four hundred thousand hippies, students, and other members of the counterculture gathered for four days of "peace, love, and music" at an outdoor concert. The concert producers had originally planned to hold the event in Woodstock, New York, but that concept fell through. They had distributed materials using the name Woodstock, however, so even though the giant gathering was held on Mr. Yasgur's farm in Bethel, the name Woodstock stuck.

No one had dreamed so many young people would actually get there to see the musical performances by artists—some very famous and some unknown. With a gathering that large, food was a huge issue, along with

People gathering at the famous Woodstock Music Festival, where granola would be launched back into popularity.

safety and sanitation. Local farm families produced an enormous quantity of food, including ten thousand sandwiches to help feed the hordes.

Hugh Romney, a Merry Prankster and member of the Hog Farm Collective, helped organize free distribution of brown rice, vegetables, and other foods. Down near the front of the stage, early arrivals had gathered and were loath to give up their positions. So Romney worked out a plan to distribute thousands of cups filled with an organic, mixed cereal made by his collective. The cups were passed hand to hand to feed the committed concertgoers.[9]

That cereal was called granola, and like Woodstock, it became a defining feature of the era. Almost as an antidote to Wonder Broad, the handmade cereal rose in popularity. It seemed new and nutritious. Granola represented a return to an environmentally sound, non-industrial food supply.

Even counterculture kitchen neophytes could make it easily in their dorm or communal kitchens, because the exact proportions of grains, seeds, nuts, spices, and dried fruits could be varied. Plus it was delicious. (A friend and I made granola in our kitchen over spring vacation. Her dad sneered, calling it birdseed, which made it taste even better to our rebellious minds.)

Little did we realize that our granola was simply a new version of the whole-grain cereals developed by Sylvester Graham and John Harvey Kellogg in the late 1800s. Like any consumer item, food has its own trends. While granola had mostly fallen out of favor between Graham and Kellogg's time and the 1960s, by the 1970s, granola trended big time. Giant food companies like Kellogg's and General Mills jumped on the bandwagon, mass-producing boxes of their own branded versions of the new cereal. It is the American way.

But don't let that stop you. Granola is still a fan favorite among straight-edged and alternative types alike. You can buy it from local providers at bakeries and food stalls or from a national brand at supermarkets, though the artisanal granolas are often the healthiest, tastiest, and most nutritious. (They lack the additives, preservatives, and food dyes that many national supermarket brands contain.) My favorite place to buy granola these days is at a small bakery called Baked and Wired in the Georgetown neighborhood of Washington, DC. They label their granola Hippie Crack and they are right—it is totally addictive. Or make it at home like we did in 1969, using the best ingredients, and it still tastes great. Listen to some Grateful Dead while you are at it.

BITE 94

❧ MESCLUN GREENS ✦

Wonder Bread versus granola. Iceberg versus mesclun greens. While foods don't face off inside a boxing ring, they do represent cultural debates in a changing society. Foods can be symbolic of a time, a place, or a person. Mesclun greens bring to mind Alice Waters, champion of fresh, local, seasonal foods and a true lover of flavorful mesclun ("mixed") baby greens.

A series of extraordinary chefs—Victoria Wise, Judy Rodgers, Jeremiah Tower, and Paul Bertolli, among others—worked at Alice Waters's legendary Chez Panisse restaurant before starting their own establishments primarily on the West Coast.

With chef Paul Aratow, she turned her Berkeley-based restaurant Chez Panisse into an organic laboratory for food rebels and gourmets when it opened in 1971. The leftist politics of that era did not dissuade her from offering the finest, freshest ingredients, which seemed a bit elitest to her radical friends. One of Waters's colleagues, Tom Luddy, observed, "As Alice used to put it, 'just because you're a revolutionary doesn't mean your idea of a good meal should be Chef Boyardee ravioli reheated in a dog dish.'"[10] Chez Panisse offered an escape from mass-marketed, humdrum food as well as from the heavily sauced formality of traditional, high-end restaurants.

Waters had studied in France, where her taste buds were awakened to delicious possibilities. She realized that great-quality vegetables, fruits, meats, cheeses, and breads could be produced in California if the American marketplace was as demanding as the French one. So she developed a network of artisanal providers. Paul Bertolli and other talented chefs, like Jeremiah Tower, used these tasty, local ingredients to create captivating meals. Each bite at Chez Panisse is a celebration of how excellent food can be. And various baby greens—young dandelion leaves, chervil, arugula, oak leaf, lamb's lettuce—make their appearance in several dishes, alone or together.

While we may welcome mesclun salad as a relatively new addition to American menus, baby greens fresh from the farm or vegetable garden are actually not a new indulgence. During his five-year assignment in Paris as minister to France, Thomas Jefferson diligently studied both French and Italian food as a scholar, a booster of American agriculture, and a gourmet. He grew eight kinds of lettuce at his breathtaking, terraced garden, dug into the hillside at Monticello, his home in Virginia.

In the early morning, an enslaved worker picked the day's lettuce leaves and returned to the kitchen. Under the watchful eye of James Hemings, Jefferson's Paris-trained chef (or any of the later cooks, all well trained), an assistant would plunge the leaves into cold water to freshen them for dinner, usually served at four o'clock. The leaves would then be tossed with an oil-and-tarragon dressing.

Jefferson adored olive oil, and tried in vain to grow olive trees he ordered as seedlings from Europe. His attempts ended in costly failure. So he generally used sesame oil on his salad instead, which his slaves pressed from the benne, or sesame, seeds grown at Monticello.[11]

More than two hundred years later, the Americans who joined Waters in taking a new, progressive look at food helped revive the demand for fresh baby greens, which proved wildly popular. By the 1990s, big supermarkets sold bags of lettuce pieces labeled as "mesclun." Yes, the bags did hold a mixture, the literal translation of mesclun, but the contents were often limp and bland. In an interview with Marian Burros, a food writer for the *New York Times*, Waters commented, "It used to be a strong, very tasty salad. Now they put things in a bag, break up big lettuces, and put in any old thing. They use it as a way to sell what they can't sell in any other form."[13]

In her Chez Panisse Menu Cookbook *(1982), Alice writes lyrically about discovering flavorful mesclun in Provence, "a mélange of the first tender young leaves which appear in the garden."[12]*

That's fine, but with all due respect to Alice Waters (who is one of my culinary heroes), my guess is that many Americans would be grateful to get access to those big bags of cut-up lettuce leaves that Waters finds so appalling. Particularly in impoverished communities, good fresh vegetables are still often few and far between.

Still, it's better than the sad reality of lettuce in the mid-twentieth century. The world has changed much for the better in its appreciation of bright, fresh greens. In the 1970s, baby leaf lettuce was very hard to find unless you grew your own or had neighbors who shared such delicacies. In a lot of America's heartland, the only lettuces in the supermarket were iceberg and sad-looking romaine until the 1990s. With the revival of farmers markets and the expanded vegetable sections in many grocery stores, our choices are much better today.

Waters has remarkably high standards and makes no apologies for them. She points to European food culture where people spend the necessary time tracking down the best, freshest ingredients and spend the necessary money purchasing those ingredients. People have found her an easy target as a food elitist.

However, Waters has worked hard encouraging young students to grow their own vegetables through her Edible Schoolyard Project, introducing young Americans from all backgrounds to the joys of fresh, seasonal food—especially mesclun greens. It's a joy we should all try to embrace.

BITE 95
GINGER CARROT SOUP

A flavorful soup and a mesclun salad together make a healthy light lunch or snack. Around the time fresh mixed greens grew in popularity, a delectably smooth, brilliantly orange soup, tangy with citrus and spice with fresh ginger, appeared on contemporary menus. Aptly named ginger carrot soup, it was simply divine and rapidly bloomed into popularity.

In cities across over the country, ginger carrot soup began to appear in all sorts of restaurants, cafés, and bistros. It was a hit in the late 1980s and '90s. Sometimes the cook made it too thick, so that the soup was more like gloppy baby food. Or the kitchen added too much orange, obliterating the carroty taste. When well-executed, however, it was a cheerful yet complex bowl of delectable healthy eating.

Around the same time, the HIV-AIDS epidemic started sweeping through the American gay community. In those early, devastating years, there was no treatment for the scourge. For straight or gay patients, HIV-AIDS was a death sentence. Meanwhile, extremists pointed to the virus as God's punishment for homosexuality. In some cases, parents who had been raised in a homophobic world suddenly had to confront the fact that their son was "queer." Some of them, often the fathers, reacted with disgust and rejection. Elton John and Bernie Taupin's beautiful work from the 1990s, "The Last Song," expresses how tragically complicated some father-son relationships were in the face of AIDS.

A BRIEF HISTORY OF GAY RIGHTS AND HIV-AIDS

Several years earlier, in 1969, the Stonewall Riot had served as a catalyst for the gay and lesbian community and their supporters to fight for their civil rights. Men and women who had long hidden their sexual orientation came out of the closet, despite the fact that homosexuality was illegal in most states. They demanded better treatment for their sick friends and partners. As the AIDS epidemic spread, protesters angrily criticized the federal government for not responding. The White House was silent. And the numbers of the sick kept growing.

By 1990, 120,453 people had died of AIDS in America. The deaths of some were covered in the news, from a young hemophiliac named Ryan White, who had been barred from his public school when he developed the disease from a blood transfusion, to Tina Chow, an avant-garde jewelry designer, to Arthur Ashe, a champion on the tennis court and in breaking down racial barriers. But most of the victims were regular people who left behind stunned, brokenhearted friends and families.[14]

The many symptoms of AIDS include weight loss and nausea. Some patients find it hard to digest animal fats but still need extra nutrients and vitamins. As the disease progresses, even chewing food can present difficulties. Ginger carrot soup is no antidote but it can be a soothing meal for someone with the disease. AIDS patients suffer from a deficiency of vitamin A, and this soup is packed with that vitamin. The ginger can help with nausea, and the smooth texture is comforting in a sore-ridden mouth.

When we feel helpless in the face of a disease like AIDS, sometimes the right food is all we can offer. It's not a remedy. But Americans often forget that natural food and especially soup can contribute to physical and emotional healing. Chicken soup has a well-deserved reputation as a cold reliever, and anemics know that calves' liver packs a powerful punch of iron. I recognize that AIDS can be fatal, placing it in an entirely different category of disease than either the common cold or most forms of anemia. Still, ginger carrot soup deserves a high rating as a soothing, nourishing food for anyone suffering from it. It's not going to cure HIV-AIDS but it can provide comfort when there is little to be found.

More than 636,000 Americans have died since the first five cases were identified in 1981. Today, thanks to dedicated scientists, adequate funding, decades of advocacy, and the invention of anti-retroviral treatments, HIV-AIDS is no longer a death sentence in the United States.[15]

GINGER CARROT SOUP

Even when you are robustly healthy, ginger carrot soup is pure, silky deliciousness. Try it as a first course for a party, or serve it alongside a fresh mesclun salad with goat cheese and toasted walnuts. You will need a blender for this recipe because the soup has to be smoothly pureed. If you have a professional-grade immersion blender, you may use that, but make sure you get all the lumps out. I'd also recommend using a Microplane grater to grate the orange peel.

INGREDIENTS

- ► 8 cups stock (chicken or vegetable)
- ► 2 medium white potatoes, peeled and chopped
- ► 2 pounds carrots, peeled and chopped
- ► 2 to 3 tablespoons grated fresh ginger
- ► 1 cup white wine
- ► 1½ cups orange juice
- ► 3 tablespoons finely grated orange peel
- ► 1 tablespoon salt or to taste

DIRECTIONS

Pour the stock into a high-sided pot. Add the potatoes. Cover and bring to a rapid boil. Add the carrots, and lower heat to medium. Cook until tender, about 15 minutes. Stir in the ginger. Remove pot from heat and let mixture cool slightly. Puree the potatoes, carrots, and stock in a blender, working in batches. Make sure each batch is completely smooth before you move on to the next. Return the puree to the pot.

Add the wine and let simmer for 10 minutes so you won't taste any alcohol. Add the orange juice and peel. Add salt. Soup should be the consistency of cream, not baby food. Thin it with more stock if needed.

Serve very hot. Top each bowl with toasted chopped hazelnuts or pepitas, or add a dollop of heavy cream.

Serves 8 to 10 as a first course or alongside a nice salad.

You may need to thin leftover or frozen soup with ½ cup stock and ½ cup orange juice.

BITE 96

QUICHE

The years of protest did not end in the 1960s with a festival highlighting peace, love, and rock music. If anything, protests grew, and a sense that the time for change had arrived motivated millions in the 1970s. First students and civil rights activists, and then a larger cross section of people protested the Vietnam War, the Nixon administration, racism, and the treatment of farm workers. The era also witnessed the rise of a new women's movement. The first wave of feminists had fought for the right to vote. In the '60s and '70s, women fought for equality in the workplace and at home. Today, historians call them second-wave feminists.

I was in school during the '70s, and grew up with the new women's movement. Writing this book, I wanted to include a food symbolic of the women's movement during this period, when activists accomplished a lot of positive change for equality that today seems normal to us. There wasn't a specific "food" for the feminist movement, per se, so I chose quiche, because it is what we ate. We ate lots of other things too, but quiche was everywhere. You could make it at home in a snap if you bought frozen pie crust (which wasn't very good back then) or you could buy it fresh at a good quality take-out store, or frozen at the supermarket, or eat it a local café. Quiche, salad, a slice of bread—it was a perfect lunch or dinner for us young women on the go. Quiche wasn't too filling, and you didn't have to eat a lot of meat, or any meat at all. Our male friends ate extra slices, and never complained.

For many Americans, quiche was their first introduction to French food. Quiche is a French word that stems from a medieval German term for baked dough. Originally, the cook mixed cream and egg with bits of smoked ham and baked the mixture in bread dough in an iron pot. It was a hearty, rustic dish. By the 1970s quiche had transformed into a custard baked in a delicate, crimped pie shell, containing various ingredients such as the traditional bacon, potato, and onion (Lorraine), spinach (Florentine), or tomato and onion (Provençal). It was not a food that carried a "Consciousness Raising" label. It wasn't organic, like the brown rice from the newly burgeoning health food stores, or associated with hippies, like granola. But by its constant presence at meetings and pot luck suppers where women talked about gender and equality, quiche fueled feminism in the 1970s.

WE WEREN'T ALWAYS SO EQUAL

The women's movement gained enough ground in the 1970s to change the lives of women throughout the United States. Influenced by the civil rights movement's quest for equality, and often united with protests against the Vietnam War, the struggle for women's equality achieved remarkable results. On August 26, 1970, tens of thousands of women across the nation demanded equal rights by participating in the Women's Strike for Equality, organized by author and feminist leader, Betty Friedan. It was the fiftieth anniversary of women's right to vote, and no one, not even leaders like Bella Abzug, Gloria Steinem, and Betty Friedan herself, knew how big the protest would be. By the end of the day, ninety marches had taken place in forty-two states. Thousands of marchers, men and women, strode down Fifth Avenue in New York City, demanding equality. Two years later, the U.S. Congress passed the legislation known today as Title IX, which guaranteed women an equal access to education among other things. From then on, high schools, colleges, universities, and medical, law, business, and other professional schools were pressured to accept women at the same levels as men and treat them equally, a revolutionary concept.[16]

In 1982, Bruce Feirstein wrote a satire entitled *Real Men Don't Eat Quiche,* in which real men watched sports, ate steak, wore big flannel shirts, and didn't share their feelings. "Quiche-eaters" or Sensitive New Age Guys, served quiche, did the dishes, and empathized with the women's movement. The parody hit home because it rang so true to many people, and Fierstein had a hit on his hands. It seemed like half the country, men and women, found feminism bewildering if not just plain wrong; after all, the Bible supported the subordinate role of women, didn't it? Quiche in many ways became the culinary symbol of the dividing line between traditionalists and progressives.

Quiche is no longer as popular as it was in the 1970s, but that doesn't make it any less satisfying. It's a snap to make, especially if you buy the crust, and it still makes everyone around the table feel equal and equally satisfied, no matter what the reality in the outside world looks like.

BITE 97
CALIFORNIA VINTAGE WINE

Nineteen seventy-six was a memorial year. Spiraling inflation damaged the economy. The Vietnam War had ended a year earlier, but news of bloodshed in Southeast Asia continued. The bicentennial of the Declaration of Independence brought an air of festivity to a nation with a somewhat battered psyche. Folks in the California wine country celebrated with a special joy.

In June, French wine connoisseurs had participated in a formal, blind tasting of similar wines from France and California. As *Time* magazine reported, "The unthinkable happened: California defeated all Gaul." In the red category, a '73 cabernet sauvignon from Stag's Leap Wine Cellars in Napa Valley surpassed all other vintage wines. The experts judged Chateau Montelena '73, a pinot chardonnay, number one in the white group. The French were stunned. The California winemakers were not surprised—they'd known how good their wine was all along—but they were over the moon with joy at the long-overdue credit. Finally, the world recognized the outstanding quality of California vintage wine.[17]

Wine making boasts a long, fascinating history. Around seven thousand years ago, humans learned how to make and store grape wine. For ancient

Babylonians, Egyptians, Israelites, Greeks, and other cultures, wine played a significant role in religious ritual, medicine, and daily life. Archaeologists have analyzed traces of wine in France from pottery dating around 450 BC. Romans spread viticulture (the growing of grapes) and viniculture (wine making) throughout their empire.[18]

Wine making in America started in the early days of European colonists. The Spanish settlers in Texas planted European grapevines, *Vitis vinifera*, near San Antonio in the early 1600s, but apparently had no lasting success. The Jesuit fathers in California proved more successful in the late 1700s. By the time the former British colonies along the East Coast had successfully won their independence, the Spanish missions had their own wine supply for dinner and the Eucharist.

World War II meant that able-bodied young men signed up for military service, leaving the grape growers short-handed. The American public, now accustomed to the sweet taste of bootlegged factory wine, had lost its palate for dry wines and looked for less sophisticated tastes.

Thomas Jefferson, who enjoyed fine wine from France and Italy, hoped to replicate the vineyards of Europe on the terraced hillsides of Monticello and personally imported French cuttings. Sadly, black rot and grape phylloxera, the tiny aphid louse, brought an untimely end to his expensive attempts at French viticulture. He turned to indigenous vines but found the results unpalatable. In Ohio, Nicholas Longworth had 1,200 acres in native Catawba grapes by 1842 and successfully produced a sparkling wine that met with great acclaim. Vineyards and wine making flourished in that area until they were wiped out by a repeating fungus, and wine making moved northeast to New York's Finger Lake region.[19]

Meanwhile, European immigrants in California—Swiss, French, and German—brought new energy to wine making on the West Coast. General Mariano Vallejo, the former Mexican commander of what is now Sonoma County, had been the first large-scale winegrower there. In the second half of the 1800s, leaders like Agoston Haraszthy and John Patchett were joined by Charles Krug, Jacob Beringer, and Carl Wente. Napa Valley and Sonoma County became wine centers for California, with excellent dry, sophisticated wines and growing businesses. In 1889, at the Exposition Universelle in Paris, the Californian wines made a grand sweep of the medals, winning two-thirds of those awarded.

Then came forty years of disease and destruction for the wine industry. The devastation began with severe frosts, followed by a plague of grape phylloxera that hit many vineyards. The 1906 San Francisco earthquake ruptured almost thirty million gallons of vintage California wine in storage, destroying years of the winemakers' craft. The most damaging event was man made. In 1920, national Prohibition took effect through the Volstead Act. Even some old-growth vineyards planted in the previous century did not escape the ax.

The proud families who had cultivated the vines and produced high-quality red wines struggled to stay afloat. They produced sacramental wines and quasi-medicinal elixirs, but this was a fraction of their former output. By the time Prohibition ended during the Roosevelt administration, a lot of the vineyards had been converted to walnut and plum orchards. It took decades, and a lot of manpower, to get the land replanted and returned to its original productivity.

Eventually, however, the winemakers produced excellent wine again and helped create a more educated wine-drinking public. By the late 1970s, California vintage wine was globally respected. The winemakers of Napa Valley and Sonoma County were only surprised by how long it took the rest of the world to catch up to what they had known all along.

Today, locations in Oregon, Washington, New York, New Mexico, and other states also produce wine acclaimed by critics and wine lovers alike. Even on the terraced hills near Jefferson's Monticello, the wine business is thriving.

BITE 98

❧ AMERICAN CHEESE ❧

It's only natural that after discussing wine, we must touch upon its eminent partner, cheese. In the last quarter of the twentieth century, cheese in America generated a number of controversies. The good-quality cheddars, mozzarella, Monterey Jack, and other traditional forms from reputable manufacturers seem to have avoided these debates. But three categories of American-made cheese—the plastic-wrapped cheese food, the handmade artisanal gourmet product, and the surplus commodity supplies—have inspired fierce commentary from die-hard fans, sneering critics, and even songwriters.

It's impossible to write about American cheese without many readers immediately thinking of the plastic-wrapped bright orange or ivory slices used in grilled cheese sandwiches and on top of burgers. Handy for sandwiches and easily melted in casseroles or macaroni dishes, these packaged goods are beloved by many, and like it or not, they play a leading role in our homegrown cuisine.

The U.S. Department of Agriculture requires that the manufacturers label these slices "process" (or "processed") cheese, "cheese product," or "cheese food" because of the additives and the factory procedures involved. Today, "singles" still define American cheese in Europe and at home, although there is so much more to American cheese than those thin-sliced concoctions. So let's strip away the cellophane and orange and yellow dyes and get to the heart of what American cheese really is.

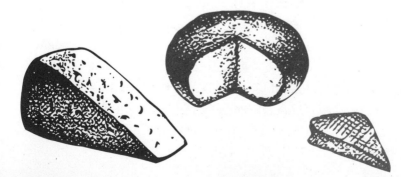

INVENTING THE BIG CHEESE

Cheese has a long history, with archaeological evidence of Neolithic production about seven thousand years ago. At some point, maybe somebody used a container made from calf's stomach to store leftover milk, and the rennet—the naturally occurring enzyme in a calf's stomach that hardens the milk solids into curds—turned the milk into cheese. (You may not personally store liquids in containers made from an animal's stomach, but they came in handy in the days before plastic bags or even glass bottles, so don't be judgmental.) The origins of the process are still murky. But the invention of cheese allowed pastoral people to preserve excess milk for later use, long before refrigeration.[20]

The Romans introduced cheese-making technology to their numerous colonies, including Hispania (Spain and Portugal). The Spanish cheese from the countryside, *queso del país*, traveled to the New World with missionaries and conquistadors in the 1500s. Wherever the Spanish settlers established dairy herds, they generally produced their own cheese. The local cheeses of eighteenth-century missions in California and Texas may have been similar to the semisoft Italian rounds and wheels beloved by the ancient Romans, who had influenced Spanish cheese making more than one thousand years earlier.

By the time of the gold rush, farmers in California produced cheese to sell to local markets, ships' chandlers, and miners. Doña Juana Cota Boronda, a mother and small business woman in Monterey, sold her family-made Spanish countryside cheese in the 1850s. Her *queso del país* may be one ancestor of the native cheese we call Monterey Jack.[21]

On the East Coast, the settlers in Plymouth, Massachusetts, imported dairy goats, and may have served fresh goat cheese along with cornmeal mush for breakfast. Different British and European immigrant groups all brought their own cheese-making traditions. Not surprisingly, the early cheese produced in New Amsterdam looked like Gouda, and the New England hard cheese resembled English cheddar. Eventually, regions with rich grasslands like Vermont and Wisconsin became major cheese producers, although farms produced cheese as a way of storing milk over the winter wherever there were dairy herds.

Factory production came early to cheese manufacturing, partially because it responded well to scale. Jesse Williams and his son opened the first cheese factory in 1851 in upstate New York, buying up local milk and producing a reliable yellow cheddar. From there, the factory system spread across the country.

It was wholesaler James L. Kraft, however, who invented American processed cheese, patented in 1916. Kraft developed this new product as a way to use up leftover pasteurized cheddar he had not been able to sell. First shredding the cheese, he repasteurized it and added sodium phosphate. The U.S. Army was an early customer on a massive scale because of the product's long shelf life. But it soon caught on with the general American public as well. Kraft's cheese became a huge commercial success, gobbling up 40 percent of the market by 1930.

Market rivals who manufactured more traditional products vehemently protested Kraft's use of the word "cheese." In response to their successful lobbying, the USDA set up guidelines for labeling processed cheese, which was legally defined as a "homogenous plastic mass" made of cheese bits and emulsifying agents, giving it predictable characteristics. In other words, when you eat Kraft "cheese," you are eating, by definition, a plastic or plastic-like product. In a short film, Kraft advertised the "meltability" and "cookability" of its cheese products, another property that its plasticity gave it.

As with the great marketing success of Wonder Bread, Kraft was able to charge more for inferior cheese by stressing the reliability and cleanliness of his invented food. Similarly, Kraft cheese was also untouched by human hands, a real selling point in that era.[22]

The 1980s saw the rise of commodity cheese. Fifty years earlier, as part of the New Deal, President Franklin Roosevelt authorized the Commodity Credit Corporation (CCC) under the USDA. It was "created to stabilize, support, and protect farm income and prices," according to the CCC home page. In the 1970s, the federal Dairy Price Support program resulted in a tremendous overabundance of cheese.

By 1981, the CCC had more than 560 million pounds of cheese in its warehouses. Only 134 million pounds were designated for the military. President Reagan authorized the distribution of this commodity cheese in December of that year in "our drive to root out waste in government and make the best possible use of our nation's resources." In the first year, 140 million pounds was distributed to needy Americans through state agencies to school lunch programs, mother and infant nutrition services, elder centers, and local food pantries. Impoverished members of tribal groups also received commodity cheese.[23]

> *The government still provides commodity cheese to state agencies for schools and other nonprofit institutions. It became highly controversial in the school lunch program, because its high fat content was in direct opposition to the government guidelines for a healthy diet.*

It took about six years to distribute all that cheese, which also was supplied in large blocks to individual families. As a result, the amount of cheese consumed in the United States skyrocketed (for obvious reasons). As more of it was put into our hands, we ate more, which then prompted us to seek it out and buy it more. Today, commodity cheese is considered a contributing factor in the obesity and diabetes epidemics because of its high fat and caloric values.[24]

The mass distribution of big blocks of cheese is clearly remembered by Americans who grew up in poverty during that era. The novelist Junot Diaz recalled removing the "government cheese" from his mother's refrigerator when a girl visited his family's apartment in his short story, "How to Date a Brown Girl, Black Girl, White Girl, or Halfie."[25] A colleague of mine at work recalled government cheese ambivalently. On the one hand, her grandmother used it to make grilled cheese sandwiches which she remembers fondly. On the other hand, she was embarrassed to have it in her apartment, because government cheese meant her family was classified as "needy."

In the blogosphere, there is an ongoing debate about government cheese. Some nostalgic recipients believe that the cheese aged to perfection in the CCC warehouses before being distributed. Policy professors urge cheese to be dropped from school menus. A rock band named the Rainmakers has recorded "Government Cheese," protesting government food supports for the needy. An artist named Keb' Mo' wrote a song in 2009, also called "Government Cheese," that empathizes with the hungry poor in America.[26]

While the CCC unloaded hundreds of millions of pounds of processed cheese and nonfat dried milk, small farmers across the nation began producing high-quality European style cheeses. Using local flavorings and innovative techniques, these artisans crafted uniquely American cheese, with evocative names like "Coach Farm's Rawstruck Wheel" and—my favorite—"Cypress Farm's Humboldt Fog."

I've heard friends sneer at these high-end cheeses, seeing them as too expensive and precious. Let's be real: this stuff isn't cheap. And it's true that fans discussing these artisanal products may sound like wine snobs pompously debating the attributes of different vintages and *terroirs*. But there's a point to that: cheese isn't all that different than wine, and thus shouldn't be treated all that differently.

CHEEZ WHIZ

Food scientists at Kraft Foods invented Cheez Whiz, a processed cheese food sauce, in 1952. It represents engineered food at its most popular. Edwin Traisman, the son of Latvian immigrants in Chicago, led the team that developed the thick, annatto-dyed sauce used on cheese steaks, hot dogs, and nachos. (Traisman also invented a method for freezing french fries that McDonald's uses.)[27]

The original Cheez Whiz was processed cheese with a list of additives, but today, even the cheese has gone, at least according to the label. "Cheez Whiz is a unique tangy flavor with zing," Kraft Foods informs buyers for food service professionals. Indeed, there is no cheese in that still-popular wide-mouthed jar, so Kraft has to spell the name of the food product with a z.[28]

When done well (and done right), it's a high-end artisanal product that blends many flavors and varies from year to year. A nicely aged, sharp cheddar from the supermarket is great to have on hand for cooking or eating with a fresh slice of a crisp apple. And while they may be more expensive than we would like, I feel thankful that the United States can now boast of its own extraordinary gourmet cheeses. There is a whole lot of deliciousness going on in this country, and our gorgeous new cheeses are part of it.

BITE 99
SALSA

It's important to recognize just what the name "salsa" really is—sauce. It doesn't mean chunky red tomatoes with tomatillos, minced onion, some capsicum, and cilantro. It just means "sauce" in Spanish. And that's what makes it so delightful—its variations are endless. The range of Latin American salsas stretches from ancient mixtures like Toltec mole to innovative combinations like Miami blueberry mango. Salsa can be tortuously hot with habanero chiles, refreshing with sage and mint, or soothing with avocado. Some commercial preparations are dull; some are exciting. Salsa, like its English translation, can be almost anything.

Salsa has evolved into a demographic symbol, a signpost of the changes in the American population and the American palate around the turn of the twenty-first century. In 1992, the *New York Times* reported that total sales of various salsas outnumbered ketchup. That's right, as you may or may not have heard, salsa surpassed ketchup as America's favorite condiment. The *Times* described this as healthy, delicious food triumphing over a traditional Anglo condiment. "The taste for salsa is as mainstream as apple pie these days," said David A. Weiss, the president of Packaged Facts Inc., a market-research company in New York.[29]

Other people were stunned that something as all-American as tomato ketchup should be overtaken in the marketplace by a Mexican food. It instantly sparked a national debate over the direction of our national cuisine. If the all-American ketchup was no longer a mainstay, what exactly *was* American food anymore? The 1990 U.S. Census showed a trajectory of population growth that seemed to mirror this finding: the country's percentage of Anglo-Americans, presumed ketchup users, was dropping.[30]

In 2007, the *Wall Street Journal* weighed in on this surprisingly popular topic by pointing out that while salsa outsold ketchup that year ($462.3 million to $298.9 million), ketchup is cheaper. So in fact, more pounds of ketchup were sold than salsa (329 million versus 184.6 million). And that wasn't even counting the billions of ketchup packets distributed at fast-food places.[31] All over America, nativists must have given a sigh of relief.

Fear of a decline in one's place in society seems to be part of the dynamic in the salsa-ketchup dispute. In truth, because we are a nation of immigrants and thus none of us (besides the American Indians) really are true-blue Americans, a portion of our population has worried about losing its majority status since the country was founded. Groups like the Know Nothings of the 1840s and the Tea Party of the early twenty-first century made anti-immigrant policies part of their political platforms.

In F. Scott Fitzgerald's *The Great Gatsby*, loutish Tom Buchanan

Some of the various ingredients in salsa.

disparages the multitudes of non-Anglo-Saxons arriving on these shores and their high birth rate. In 1920, eight hundred thousand people immigrated to the United States, and Buchanan's criticisms in the novel reflected actual concerns during the period in which Fitzgerald was writing. (To provide more context, the peak year for immigration in U.S. history was 1907, with more than 1.2 million legal arrivals. In 2010 there were two hundred thousand fewer than 1907, but that perspective gets lost in the immigration debates. People's fears about being overrun by illegal immigrants seem widely out of balance with today's figures. So much of the ketchup-versus-salsa debate is simply shorthand for concerns about changing cultural dominancy.)

The truth about salsa purchasers is that most of them are also ketchup purchasers.[32] People use the two products differently and happily consume them with different foods. I love fresh pico de gallo with a breakfast burrito but swear by ketchup with my hamburger, and I'm not switching. Apparently, most Americans agree. There is room for both in our refrigerators, and our national gastronomy is richer for it.

MANGO SALSA

Lots of people have a favorite salsa, regardless of their ancestry. Mine is chopped fresh mango mixed with liberal amounts of minced cilantro, 2 tablespoons of minced red onion, juice from one small lime, and a touch of that fabulous all-American Sriracha hot sauce to give it a kick. It may not be authentically Latin, but it is authentically good. From my point of view, the future of our cuisine lies in fusion.

Serve it with grilled duck breast or a swordfish steak. Grill either one medium rare on a summer night. Delicioso!

BITE 100

☘ SUSHI ☘

In the 1970s, sushi was an exotic food, consumed almost solely by members of the small Japanese communities and Japanese business travelers in California and New York City. Westerners in these restaurants were generally either business clients of Japanese industries or among the most adventurous eaters.

Meanwhile, Japan's economy, fully recovered from the devastations of World War II, powered into global prominence. Young sushi chefs came to the United States to open new restaurants, an expensive, slow, and byzantine process in Tokyo.[33]

> *During the sushi boom in the late twentieth century, bluefin tuna, a big coastal fish sold for cat food a decade earlier, became a new luxury item.*[34]

Although it was a new gastronomic experience in America, sushi had a long history in Asia. Archaeologists have found 2,500-year-old prototypes of sushi in China along the Mekong River. By the 1800s in Edo (today's Tokyo), vendors shaped vinegared rice and topped it with very fresh raw fish, selling the same version of sushi we would recognize today.

By the early 1980s, sushi restaurants began to dot America's big cities. The gourmet advance guard embraced the novel tastes, smells, textures, and presentation of this exotic food from the Far East. For example, in the 1985 cult movie *Breakfast Club*, rich girl Claire, played by Molly Ringwald, brings sushi for lunch, which adds to her princess reputation. Sushi was an object of desire for some with its delicate texture, and an object of disgust for others: "Raw fish, yuck!" Despite the negative initial response, sushi bars proliferated, their number quintupling between 1988 and 1998. Chefs invented American variations, like the California roll, with avocado and amalgamated crab stick, and the Philadelphia roll with cream cheese, which Japanese and American traditionalists found disconcerting.[35]

Other foreign restaurants blossomed, often following various debacles of failed American foreign policy. In communities where Americanized

Chinese and Tex-Mex once represented the exotic, Thai and Vietnamese takeout places opened, joined by Middle Eastern tavernas, Indian curry diners, Korean barbecues, and Latin Caribbean kitchens. Now a food adventurer could find a cornucopia of international cuisine in New York, Chicago, Los Angeles, and Houston.

As the new millennium approached, residents in small towns, as well as big cities, had no need to read demographic statistics or immigration policy reports that analyzed the new immigrants. They could figure out the significant presence of new immigrant groups simply by reading the Yellow Pages under "restaurants." Aromas from the four corners of the earth scented the nation's Main Street. People from all over the world were coming to America for new opportunities. And some of them were setting up restaurants to showcase and share the cuisines from their old countries with the people of their new homelands, just like the Italians, Germans, Greeks, and Jews did more than one hundred years ago.

Food trends come and go. Americans no longer serve guests syllabub in special glassware as we did in the eighteenth century—if they even serve it at all. Most people prefer lobster to eel these days, unlike the *Mayflower* arrivals. The beverage powder Tang lost its luster fairly quickly. Our elders still bring Jell-O with pride to gatherings, perplexing millennial grandchildren who don't understand the appeal, at least without vodka.

It's hard to foresee which foods that we perceive as exotic today will be woven into the permanent culinary traditions of the United States, while others we take for granted may turn out to be flashes in the pan. Who knows how much longer raw fish on rice will be popular or salsa the "new ketchup"? However, we can be sure that emblematic foods from other countries will continue to join our national menu as "new," just as people from other countries continue to enrich our own with their experiences, tastes, creativity, and traditions. Other foods will fade into history. We are a land of immigration and innovation, and our diversity makes us stronger. This is not a static country. As America continues to grow and change, our food will too. Our cuisine responds to the shifting rise and fall of culture, taste, and desire.

A FEW EXTRA BITES

American Food Today

As a historian, I believe that time and experience bring a perspective that is missing when you are in the middle of historical events as they unfold. So I am hesitant to talk too much about twenty-first-century foods and food trends, since at the time I'm writing this, we simply don't have that perspective yet. There are, however, a few stories about food that illustrate some early twenty-first-century developments, providing what I hope is a flavorful end to this book.

✿ CHILI CON CARNE ✿

Some people may think it is shocking that I left chili con carne to the epilogue. After all, bowls of chili-spiced beans, with or without meat, surfaced early in the history of the American Southwest and they've become integral to our national cuisine. (Just look at the annual chili mania that occurs around the Super Bowl each year.) A local food in Texas and neighboring areas, chili spread north, west, and east after the Mexican War during President James Polk's expansionist administration.

Like a spicy mirror of the United States in the late 1840s, chili had its own sense of manifest destiny, spreading from sea to shining sea. Suppliers for cattle drives created chili bricks, rectangles of dried beef, beans, suet, minced chilies, and other seasonings that could be reconstituted with water for a quick and easy meal on the trail. In this way, chili bricks resembled a nineteenth-century version of American Indian pemmican: they were portable and packed a significant amount of nourishment and calories. Then the invention of commercially available powdered chili brought various chili recipes into home kitchens that had never seen a hot red pepper before. After the 1893 World's Columbian Exposition in Chicago, where thousands of visitors enjoyed chili for the first time, Americans ate chili with gusto.[1]

Like barbecue, there are hundreds of recipes for chili con carne today, each with its own legions of supporters. Cable TV shows thrive on chili how-to's, competitions, and other culinary explorations. Food bloggers and recipe sites can debate chili methodology endlessly. Texans swear by an all-beef, hot-spiced "bowl o' red," while Springfield, Illinois, which claims to be the "chilli" capital, offers a milder dish in which kidney beans share the pot with ground meat. In Cincinnati, Ohio, folks proudly pour their Skyline chili over a plate of spaghetti, almost like a meat sauce. Served up with salad, corn bread, and maybe some rice, it makes a perfect one-pot meal for football fans or anyone who needs to feed a large crowd over an extended period of time.

A big pot of simmering chili fits right into the needs of a firehouse kitchen as well. Filled with hardworking firefighters who come and go around the clock and in all weather, firehouses produce stews and hearty chowders along with chili for their hungry crews, dishes that can simmer for hours without losing their flavor and then be ready to serve at a moment's notice. In fact, chili's flavors often improve over time as the spices and other ingredients have a chance to mingle.

New York City has many historic firehouses, some of which boast terrific cooks as well as brave firefighters. My favorite firehouse, home to Engine Company 33 and Ladder Company 9, is on Great Jones Street, right off the Bowery on Manhattan's Lower East Side. A stunning Beaux Arts building constructed in 1899, it stands in testimony to the era when municipal struc- tures reflected the grand hopes and aspira- tions of a generation. But the neighborhood slid downhill, becoming a dirty and danger- ous place for much of the twentieth century. So many heroin addicts lived in the vicin- ity that urban legend pins the derivation of "jonesing," a synonym for a bad craving, on the street's name.[2]

Things began to turn around for that neighborhood in the 1990s, when gentrifi- cation turned the shadowy edges of lower Manhattan into prime real estate. By the beginning of the new century, the Bowery and its side streets housed new restau- rants, shops, and restored apartment build- ings. Meanwhile, Engine Company 33 and Ladder Company 9 thrived—until a bright, sunny September morning in 2001.

The firehouse home of Engine Company 33 and Ladder Company 9 on Great Jones Street in Manhattan.

The terrorist attack by al Qaeda oper- atives on the World Trade Towers claimed more than three thousand lives that September 11, 2001. This single event altered the course of U.S. history, and it changed lives as well. Ten firemen from the firehouse on Great Jones Street are numbered among those lost, more than from any other single firehouse in New York City. That day left a scar that will never truly heal for that brave crew of firefighters, their loved ones, and their communities. A bronze plaque on the brick front lists their names. We all know a plaque will never be enough.

I visited the firehouse on Great Jones Street in 2014 and met Lieutenant Bryan Gomoka and a team of terrific firefighters from Engine Company 33 and Ladder Company 9. They proudly serve New York City twenty-four hours a day, seven days a week. Like their predecessors in the early twentieth century, they still find comfort in gathering over hot chili simmering in a big pot and ladled out in individual bowls for firefighters whenever they return to Great Jones Street after a call.

Here is the recipe that Lieutenant Gomoka shared with me, which I am proud to share with you.

FIREHOUSE CHILI CON CARNE

Bryan Gomoka said that several of the firefighters at the Great Jones Street firehouse have contributed to this recipe. They have a great kitchen at the firehouse and some excellent cooks too. Here is the original recipe. I guess the firehouse likes its chili super spicy, or Bryan was teasing me, because ½ cup chili powder seems like a boatload to me. I would start with 2 to 3 tablespoons of chili powder and build up from there. Otherwise, this recipe definitely qualifies as a four-alarm fire.

INGREDIENTS

- ► 2 pounds ground beef (or turkey or bison)
- ► 1 green pepper
- ► 1 head garlic
- ► 4 (14½-ounce) cans diced tomatoes
- ► 2 (14½-ounce) cans kidney beans

- ▶ ½ cup chili powder
- ▶ 2 tablespoons ground cumin
- ▶ ¼ teaspoon dried thyme
- ▶ Salt, freshly ground black pepper, and cayenne pepper to taste

DIRECTIONS

Brown the meat in a large frying pan. Chop the green pepper and peeled garlic cloves. Put the meat, green pepper, garlic, tomatoes, beans, chili powder, cumin, thyme, salt, and pepper in a big pot and cook over low heat for 2 to 3 hours, covered. Check on this and stir once in a while.

Serves 8 to 10.

You also can add 2 minced jalapeños. Remove the seeds, or leave them in if you like chili even more hotly spiced.

BITE 102

✹ SUPER FOODS ✹
AND DIETS

Chili boasts a long history, and while people add and subtract from the basic recipe, it has remained pretty much the same spicy, slow-cooking dish for more than 150 years. The concept of super foods is the exact opposite. These days, the top picks seem to change every month. Some of that is due to the Web, where new ideas, theories, and crazy suggestions can reach an audience instantly.

Though "super foods" is a relatively new term, every era has had its own version of them—foods that steal the spotlight briefly because they're considered remarkably nourishing and intrinsically good for you. Throughout the colonial period, Americans consumed the super foods of beef, pork, and wild game. These were considered the mainstays of a healthy diet, along with fruit pies. Nineteenth-century food reformer Sylvester Graham advocated the super-food power of whole grain bread, biscuits, and cereal, and attracted dedicated followers, known as Grahamites, who believed that whole grain foods and vegetarianism would make the world a better, healthier place. (The two trends are making a comeback today.)

In the 1980s and early 1990s, fresh pasta hit the big time, along with sun-dried tomatoes, as Americans ardently embraced Italian culinary culture as an ideal antidote—low-fat!—to the rise of fast food. (Until the Atkins Diet and the low-carbohydrate diet came along, that is.) In the twenty-first century, the most popular super foods are touted for their supposed ability to promote anti-aging and an overall healthier, longer life. Acai ranks high on several lists, as do blueberries, tofu, wheat germ, green tea, and yogurt.

Let's be real here: "super food" is a contemporary marketing term. There's no scientific definition for it; no parameters for what it should (or shouldn't) do for our health and how it can be used (or not) on food labels; and no means to test and evaluate the effectiveness of its culinary superpowers. That doesn't stop nutritionists, dietitians, and the media from issuing lists of the top super foods and dueling over which ones really qualify for this arbitrary title. For example, Oprah Winfrey's O magazine's website features ten super foods picked by Nicholas Perricone, MD, because they're "rich in essential fatty acids, antioxidants, fiber, or all three!" Mehmet Oz, MD, host of the Dr. Oz Show, recommends seaweed because it's the "new super food for immunity."[3]

Meanwhile, across the Atlantic, the European Union has banned the use of the word "super food" on product labels "unless accompanied by a specific medical claim supported by credible scientific research." An essay in the British newspaper the Guardian quotes Catherine Collins, the chief dietitian at St. George's Hospital, who speaks frankly on the topic. "The term 'super foods' is at best meaningless and at worst harmful. There are so many wrongful ideas about super foods that I don't know where to begin to unpack it."[4]

The very concept of super foods embodies my favorite response to that old adage, "History repeats itself." Throughout history, humans have lurched from one dietary fad to the next, convinced in one moment that oysters serve as a potent aphrodisiac and in the next that eating whole grain bread will end violence. But this is not the past reoccurring over and over. History does not repeat itself. It just proves that human nature remains the same. A great example of this is the idea of super foods. In a society where most people have the freedom to choose what they want to eat and how much, a significant portion of them will always be searching for answers to life's questions or issues on their plate. Interestingly, the term super foods appears to be trending down in the spring of 2014, but the promise of health and beauty from a few foods continues in magazines and websites.

Like super foods, certain diets promising health and longevity have been on the rise. Some of these special diets are increasing in popularity as our awareness and ability to diagnose food allergies and intolerances improves. An individual with celiac disease, for example, requires a gluten-free diet. Others simply prefer to avoid gluten, a naturally occurring protein composite found in many grain-based foods, especially breads and pastas, for the very legitimate reason that doing so makes them feel better.

Because of this rising demand, today a consumer can locate gluten-free products in traditional supermarkets and delis as well as health food stores. Even in Italy, restaurants often feature gluten-free pasta on their menus nowadays. Ten years ago, gluten-free pasta was hard to find anywhere.

A gluten-free diet is much less demanding than another kind of special diet, a vegan regime, which avoids all animal products—meat, fish, eggs, dairy, and so on—and sometimes processed foods too. Veganism, like vegetarianism, often incorporates a personal stance against taking, killing, or mistreating animals. Strict gastronomical rules also appear in certain fundamentalist or conservative religious communities. Some of these were originally based on overt health issues from centuries ago, such as the banning of pork as an unclean meat by Semitic groups.

On the other end of lifestyle dietary trends are fad diets often promising weight loss in ways that seem bizarre once their moment in the sun fades. In 1925, Lucky Strike cigarettes promoted their brand through a diet campaign that advised: "Reach for a Lucky instead of a sweet." It seems odd (and almost horrifying) today that a cigarette company would have touted its products as the "healthier" option for consumers. A few years later, the popular Hollywood Diet was born, based on eating grapefruit with every meal. The Cabbage Soup Diet caught on in the 1950s and reappeared in the early twenty-first century in a modified, French form in the 2004 bestseller *French Women Don't Get Fat*, which encouraged the consumption of leek soup.[5] Weight-loss diets appear laughable with the perspective time grants us.

With that historical hindsight, I still particularly enjoy seeing blueberries on many of the super food lists. As we saw back in Chapter 1, long before the arrival of the Europeans in the New World, our American Indian forerunners had a good sense of what was nourishing. The nutritional punch of blueberries was certainly appreciated then as it still is today.

Another of my favorite healthy choices is tofu, a lean, versatile protein based on soy beans. Below is a bit of history that may surprise you.

A HISTORY OF TOFU

Tofu has a long, impressive history in China. As early as 1600 BC, the ancient Chinese cultivated soybeans, which remain a vital source of protein and other nutrients throughout Asia today. During the Han Dynasty (between 206 BC and 220 AD, about the same time as the Roman Empire's glory days), the Chinese production of bean curd became widespread and was called "doufu." Eventually, Buddhist monks brought soybeans and recipes for bean curd to Japan, where it was adopted and rechristened "tofu," the term many Americans use for it today.

It took a long time for tofu to come to the United States. An early advocate was Samuel Bowen, a well-traveled sailor who settled down near Savannah, Georgia, and planted soybeans for his employer in 1765. The approach of the American Revolution stymied this enterprise.

But a few years later, tofu found another champion in the early days of the new American republic: a man remembered as a scientist, diplomat, writer, editor, postmaster, signer of the Constitution, and a true character who liked to take air baths naked. That's right—Ben Franklin. In 1770, Franklin sent soybeans (which he called "Chinese caravances") back from France to his friend John Bertram, a famous botanist, exclaiming in his accompanying letter about a "special cheese" made from the beans "which is called Tau Fu."

I first encountered this story of Franklin's appreciation for tofu at the City Tavern, a historic restaurant in Philadelphia that offered a version of the dish inspired by this Founding Father. I could see why Ben Franklin loved it. My grilled bean-curd entrée was *very* tasty.

BITE 103

✥ MOLECULARLY ✥ MODIFIED FOODS

Super foods illustrate Americans' passion for anything new on the dietary front, but they are just one trend in today's world. Two other gastronomic trends involve altering the molecular structure of food through science in polar opposite ways.[6] One is molecular gastronomy, a contemporary form of cuisine that alters the structure of food through organic chemical and temperature applications. It represents science applied to foods in the professional or home kitchen, and creates an almost avant-garde culinary experience.

The other is the genetic modification of food crops and animals, in which the DNA of the organism is altered to improve, enrich, or enhance crop production, human health, or other qualities. GMOs (genetically modified organisms) represent bioengineering of organisms in the laboratory for use in agriculture.

Molecular gastronomy produces highly flavorful dishes, often with surprising textures and appearances. For example, you might encounter delicate pink foam on top of a salmon slice in a restaurant. That may be foamed beetroot, created through the introduction of flavorless soy lecithin. Or little bright-green rounds of what looks like fish eggs may be mint syrup "caviar" on top of your ice cream, produced through a "spherefication" process using sodium alginate. Some of the most inventive and highly regarded cooks in today's world of celebrity chefs have embraced these new techniques, producing surprisingly intense and delicious results that may seem bizarre to our eyes but wonderful to our taste buds.[7]

Individuals like Spain's Ferran Adria, head chef at the (now closed) three-star restaurant El Bulli, and Grant Achatz at Chicago's Alinea have explored food and cooking innovations like these, delighting some and perplexing others. In an age full of farm-to-table enthusiasts, the dripping of mint syrup into a sodium alginate bath with an eyedropper strikes some critics as a pursuit more appropriate to the lab than the kitchen. Even the label "molecular gastronomy" provokes intense debate; it's considered too elitist or a virtual food by some. Celebrity chef Anthony Bourdain remarked, "It doesn't sound like anything I would be in the mood for...but when you complain [as a chef], you're like my grandparents complaining about the electric guitar."[8]

All of this debate about new molecular gastronomy techniques seems like a minor kerfuffle, however, compared to the arguments raging domestically and internationally over GMOs, also known as genetically engineered food.

Let's be honest. For millennia, farmers have sought specific traits in their crops and animals, crossbreeding and cross-pollinating for taste, sturdiness, and yield. George Washington personified the scientific farmer of the Enlightenment, carefully breeding his cattle as well as his foxhounds to obtain the finest specimens. And in the nineteenth century, American farmers created more than a thousand varieties of apple trees through grafting, in search of the ideal cider, baking, storing, and eating apples.[9]

Genetic engineering is different, because the process introduces genes from another organism directly into the host's DNA. For example, in 1995, Monsanto Corporation introduced a sturdier tomato that could ship without bruising by engineering, or restructuring its chromosomes. Monsanto hoped to increase yields for farmers and its own seed sales, both reasonable goals in and of themselves. Critics of GMOs, however, argue that the new, artificially created tomato might breed with other natural tomatoes via cross-pollination, and there would be no way to keep the engineered DNA from escaping out into the world. The new, so-called "Frankenstein" tomato failed.

A tougher issue is raised when a GMO food might have the capacity to save lives—in this case, about 1.5 million lives a year. Two scientists invented the prototype Golden Rice in 2000, rice genetically modified to include vitamin A (beta-carotene) by introducing genes from daffodils and a specific bacterium. In the developing world today, a deficiency of dietary vitamin A kills men, women, and especially young children and blinds an additional five hundred thousand people *every year*. By switching to Golden Rice, these lives could be saved and the quality of life for entire communities improved, according to proponents.[10]

The debate over Golden Rice rages on. Greenpeace, the international environmental group, is adamantly opposed to all GMOs, observing that we can't foresee the dangers that bioengineers may be unleashing. On the other hand, in 2013 the new Pope Francis gave his personal blessing to Golden Rice 2, with the caveat that the poor should be the primary beneficiaries, not big business.

Michael Pollan, American food writer and sage, believes that more research is needed. Groups like UNICEF and the World Health Organization advocate the efficient distribution of vitamin A supplements and the encouragement of local families and small farmers to grow more carrots, sweet potatoes, and other foods that provide natural sources of vitamin A.[11] There is no easy answer to this controversy.

Why am I even bringing up an international food policy debate when this book is about the history of the United States? Because it is a perfect example of where America is today in our relationship to food. We have moved from being rustic colonies of Spain and Britain to a restless and expanding young nation to a world superpower. From roast beaver tail to mesclun salad, pemmican to TV dinners, and blueberries to, well, blueberries, our culinary choices parallel the stages of our historical development, and they've helped us grow into a vital producer, exporter, and importer of foodstuffs. So it should not be a surprise that our actions on this front also have an international impact.

As we continue to become increasingly interconnected with the rest of the world, global questions about diet and food resources inevitably will involve our country. As a nation, the choices we make about the food we eat and the water we drink no longer exist in isolation, if indeed they ever did. Adequate nutrition, the sustainability of food crops, and the preservation of historic seeds are vitally important for everyone. Our food decisions impact not only our own lives, but the lives of people everywhere and for generations to come. Along the way, we should remember to preserve the delicious, edible heritage that we share today on the American plate.

☙ END NOTES ☙

CHAPTER 1

1. Linda Civitello, *Cuisine and Culture, a History of Food and People*, 3rd ed. (Hoboken, NJ: Wiley, 2011), 104, 109; Joseph Stromberg, "Ancient Popcorn Unearthed in Peru." Smithsonian.com (January 27, 2012) www.smithsonianmag.com /smithsonian-institution/ancient-popcorn-unearthed-in-peru-81304339/?no-ist (accessed December 8, 2013). The selective breeding of maize has continued for four hundred years, resulting in the varieties we recognize today at the farm stand and at the feed store.

2. Richard Hetzler, *The Mitsitam Café Cookbook: Recipes from the Smithsonian National Museum of the American Indian* (Washington, DC: Smithsonian Institution, 2010), 139.

3. Joseph E. Dabney, *Smokehouse Ham, Spoon Bread, and Scuppernong Wine: The Folklore and Art of Southern Appalachian Cooking* (Naperville, IL: Cumberland House Publishing, 2010), 323–25.

4. Civitello, *Cuisine and Culture*, 113.

5. Civitello, *Cuisine and Culture*, 100.

6. C. W. Severinghaus and C. P. Brown, "History of the White-Tailed Deer in New York," *New York Fish and Game Journal* 3, no. 2 (July 1956): 8–11.

7. To more fully understand the complex interactions between Europe and the New World generally, see Jared Diamond, *Guns, Germs, and Steel* (New York: W. W. Norton, 1997).

8. First Peoples Buffalo Jump State Park, stateparks.mt.gov/first-peoples-buffalo-jump/.

9. A brief description of the policy of buffalo slaughter can be found at www.pbs.org /buffalowar/buffalo.html.

10. Edmund Jefferson Danziger Jr., *The Chippewas of Lake Superior* (Norman: University of Oklahoma Press, 2012), 12. See also University of Vermont, "Maple Syrup: History of Maple," library.uvm.edu/maple/history/.

11. Lolita Taylor, "Objiwa, The Wild Rice People," (unpublished paper, 1973) cited by Edmund Jefferson Danziger Jr., *The Chippewas of Lake Superior*, 13.

12. E. Barrie Kavasch, *Native Harvests: American Indian Wild Food and Recipes* (Mineola, NY: Dover Publications, 2005), 50.

13. Encyclopedia of Life, "Capsicum Peppers," eol.org/pages/38876/overview.

14. Bill Laws, *Fifty Plants that Changed the Course of History* (Richmond Hill, ON: Firefly Books, 2011), 38.

15. University of Washington Libraries, "Salmon in the Northwest," from University of Washington Libraries, Digital Collections website (2012), content.lib.washington .edu/salmonweb/index.html (accessed October 26, 2012).

CHAPTER 2

1. Dave DeWitt, *The Founding Foodies: How Washington, Jefferson, and Franklin Revolutionized American Cuisine* (Naperville, IL: Sourcebooks, 2010), 26–31.

2. Jane C. Nylander, *Our Own Snug Fireside: Images of the New England Home, 1760–1860* (New Haven, CT: Yale University Press, 1994), 205–6.

3. James E. McWilliams, *A Revolution in Eating: How the Quest for Food Shaped America* (New York: Columbia University Press, 2005), 2–3.

4. McWilliams, *A Revolution in Eating*, 80–82.

5. Fynes Moryson's multivolume *Itinerary* (1598) has been digitized and can be viewed online at several websites. He mentions cockaleekie soup with prunes being served at a knight's house for dinner. See articles.sun-sentinel.com/1986-01-22/features/8601050761_1 _burns-night-haggis-burns-supper.

6. McWilliams, *A Revolution in Eating*, 252–53.

7. Reay Tannahill, *Food in History* (New York: Three Rivers Press, 1988), 210–11.

8. Nylander, *Our Own Snug Fireside*, 226–28.

CHAPTER 3

1. Stephen McLeod, ed., *Dining with the Washingtons* (Chapel Hill: University of North Carolina Press, 2011), 200–201.

2. Libby H. O'Connell, "Lunch with Libby: A Founding Father's Favorite Food." www.history .com/news/hungry-history (accessed May 13, 2014).

3. Peter G. Rose has researched and written extensively on this era in New Netherland. Peter G. Rose, trans. and ed., *The Sensible Cook: Dutch Foodways in the Old World and the New* (Syracuse, NY: Syracuse University Press, 1998), 26–27, 29, 76–77.

4. "11 Facts about American Eating Habits." www.beta.dosomething.org (accessed April 30, 2014), citing Sally Levitt Steinberg, *The Donut Book* (New York: Workman Publishing, 2004).

5. T. H. Breen, "An Empire of Goods: The Anglicization of Colonial America, 1690–1776." *Journal of British Studies* 25, no. 4 (October 1986): 467–99.

6. Ed Crews, "Rattle Skull, Stonewall, Bogus, Blackstrap, Bombo, Mimbo, Whistle Belly, Syllabub, Sling, Toddy, and Flip: Drinking in Colonial America." *Colonial Williamsburg Journal* Holiday, 2007. www.history.org/Foundation/journal/Holiday07/drink.cfm.

7. Light Horse Harry Lee, an early American patriot and Revolutionary War officer, famously spoke of George Washington at the latter's funeral as "First in war, first in peace, and first in the hearts of his countrymen." Carrie Shanafelt, "Item of the Day: Funeral Oration for George Washington by Major-General Henry Lee (1800)," *Eighteenth-Century Reading Room* (blog), May 22, 2006. 18thcenturyreadingroom.wordpress .com/2006/05/22/item-of-the-day-funeral-oration-for-george-washington-by-major-general-henry-lee-1800/ (accessed July 11, 2014). See also ibid.

8. Peter J. Hatch, *"A Rich Spot of Earth": Thomas Jefferson's Revolutionary Garden at Monticello* (New Haven, CT: Yale University Press, 2012), 3–10. This beautiful book is also a fascinating history of one man's garden.

9. Mary Randolph, *The Virginia Housewife, or, Methodical Cook* (1824; facsimile ed., Gloucester, England: Dodo Press, 2007), 76.

10. Libby H. O'Connell, "Lunch with Libby: the History of Peas," www.history.com /hungryhistory (accessed May 8, 2014).

Chapter 4

1. Frederick Philip Stief, *Eat, Drink, and Be Merry in Maryland* (New York: G. P. Putnam and Sons, 1932), 31–35. Thanks to my dear friend Peach Schnier for lending me this family heirloom and rich resource.

2. *William Black's Journal, 1744*, excerpted in Linda Stradley, "History and Legends of Ice Cream and Ices," whatscookingamerica.net/History/IceCream/IceCreamHistory.htm (accessed March 26, 2014).

3. Joseph E. Dabney, *Smokehouse Ham, Spoon Bread, and Scuppernong Wine: The Folklore and Art of Southern Appalachian Cooking* (Naperville, IL: Cumberland House, 2010), 216.

4. Later in the antebellum period, pearl ash became commercially available, but yeast cakes remained popular. Pearl ash could have an alkaline flavor that was unwanted in a delicate cake, although Amelia Simmons used it in a spicy gingerbread recipe. See Louise Conway Belden, *The Festive Tradition: Table Decoration and Desserts in America, 1650– 1900* (New York: W. W. Norton, 1983), 182–83. On the history of baking powder, see Linda Stradley, "History of Baking Powder," whatscookingamerica.net/History /BakingPowderHistory.htm.

5. Thanks to Mary V. Thompson, research historian, and Rob Shenk, vice president, both at George Washington's Mount Vernon, who shared this information with the author, January 14, 2014. Hercules, who was formally trained as a chef, eventually ran away to freedom when the Washingtons were living in Philadelphia.

6. Andew F. Smith, "Stoves and Ovens," in *The Oxford Encyclopedia of Food and Drink in America*, Vol. 2 (New York: Oxford University Press, 2012), 499–500. Stewart based his Oberlin stove on the work of one of America's great practical scientists and inventors, Benjamin Thompson, Count Rumford.

7. Amelia Simmons, *American Cookery* (Hartford, CT: 1796; facsimile ed., Mineloa, NY: Dover Press, 1984); Randolph, *The Virginia Housewife*, 60. Mary Randolph was married to a cousin of Thomas Jefferson.

8. Dan Strehl, ed. and trans., *Encarnación's Kitchen: Mexican Recipes from Nineteenth-Century California, Selections from Encarnación Pinedo's* El cocinero español (Berkeley: University of California Press, 2005), 108.

9. Andrew Beahrs, *Twain's Feast: Searching for America's Lost Foods in the Steps of Samuel Clemens* (New York: Penguin Books, 2011), 214. This is a great book.

10. Report in the *Cork Examiner*, 1846, cited in www.irishhistorian.com/IrishFamineTimeline .html.

11. Gerald A. Danzer et al., *The Americans* (Boston: McDougal Littell, 2009), 264.

12. Simon Ford, "What Exactly Is a Mint Julep? And Why Should I Be Drinking It All Year Long?" *Food Republic*, May 2, 2013, from www.FoodRepublic.com (accessed March 23, 2014).

13. Jessica Harris, *High on the Hog: A Culinary Journey from Africa to America* (New York: Bloomsbury, 2011), 97, 152. Slaves also supplemented their diet with game whenever possible.

14. Thank you, friend from Tennessee, and others, for this childhood memory. "You couldn't eat the chitlins prepared by some mothers."

CHAPTER 5

1. Nancy F. Cott, *The Bonds of Womanhood* (New Haven, CT: Yale University Press, 1977), 64–75; Quotation from Caleb Cushing, "The Social Condition of Woman," from the *North American Review*, April 1836, reprinted in Allen Thorndike Rice, ed., *Essays from the North American Review* (New York: D. Appleton, 1879), 67.

2. Abraham Lincoln, quoted in "Mary Todd Lincoln." www.whitehouse.gov/about/first-ladies /marylincoln (accessed May 3, 2014).

3. Donna F. McCreary, *Lincoln's Table* (Zionsville: Guild Press of Indiana, 2000), 57.

4. William C. Davis, *A Taste for War: The Culinary History of the Blue and the Gray* (Lincoln: University of Nebraska Press, 2003), 22, 111, 122. The food depot at City Point, Virginia, was a remarkable wartime accomplishment. See "Petersburg National Battlefield: City Point," www.cr.nps.gov/logcabin/html/cp.html.

5. John D. Billings, *Hardtack and Coffee: The Unwritten Story of Army Life* (1881, facsimile ed., Lincoln, NB: Bison Books, 1993), 113–22.

6. Lincoln noted the beneficial weather conditions in his announcement of the first national Thanksgiving in the fall of 1863.

7. Andrew F. Smith, *Starving the South: How the North Won the Civil War* (New York: St. Martin's Press, 2011), 49–55.

8. Libby O'Connell, "Giving Thanks: At Home and in the Field, the Civil War Wrecked Havoc on a New Holiday." *The History Channel Magazine*, November/December, 2011, 44–50.

9. Ibid.

10. Smith, *Starving the South*, 145–62.

11. PBS, "Workers of the Central Pacific Railroad," *American Experience*, www.pbs.org/wgbh /americanexperience/features/general-article/tcrr-cprr; Rhonda Parkinson, "Chop Suey: A Cultural History of Chinese Food in the United States," book review, at "Chinese Food," About.com, chinesefood.about.com/od/diningout/fr/coe-book-review.htm (accessed January 17, 2014).

12. David R. Chan, "Menuism: How American Chinese Food Came to Be," *Huffington Post Blog*, September 26, 2012, www.huffingtonpost.com/Menuism/how-american-chinese-food-came-to-be_b_1902395.html (accessed January 17, 2014).

13. Carolyn Hughes Crowley, "The Man Who Invented Elsie, the Borden Cow," *Smithsonian*, September 1, 1999, www.smithsonianmag.com/science-nature/the-man-who-invented-elsie-the-borden-cow-171931492/?no-ist (accessed January 14, 2014).

14. The 2010 U.S. Census reports that people of German descent continue to make up the largest single ancestry category in the United States, with 15 percent of the population (47.9 million) claiming German ancestors. The peak year of immigration was just before the outbreak of the Civil War, in 1859, when almost one million Germans arrived in America. See U.S. Census Bureau, "American FactFinder: People Reporting

Ancestry, 2010," factfinder2.census.gov/faces/tableservices/jsf/pages/productview .xhtml?pid=ACS_10_1YR_B04006&prodType=table.

15. The Zerega pasta company exemplifies the classic American immigration story. See "History: An American Original," at Zerega, www.zerega.com/history.html. Thanks to my dear friends, the Vermylen-Zerega family for sharing their story.

16. David Paccioli, "Dark Legacy," *Penn State News*, May 2004. news.psu.edu/story /140775/2004/05/01/research/dark-legacy.

17. Hasia R. Diner, *Hungering for America: Italian, Irish, and Jewish Foodways in the Age of Migration* (Cambridge, MA: Harvard University Press, 2002), Loc 822 of 3746.

18. Matthew S. Magda, "The Welsh in Pennsylvania: The Peoples of Pennsylvania." Pennsylvania Historical and Museum Commission, 1998, www.portal.state.pa.us /portal/server.pt/community/groups/4286/welsh/471948 (accessed January 17, 2014). See also Dave Tabler, "Every Coal Miner's Lunch Smelled of the Coal Mines," Appalachian History, June 19, 2008, www.appalachianhistory.net/2008/06/every-coal-miners-lunch-bucket-smelled.html.

19. Lisa Bramen, "The History of the Lunch Box," *Smithsonian*, August 31, 2012, www .smithsonianmag.com/arts-culture/the-history-of-the-lunch-box-98329938/ (accessed March 23, 2014).

20. Ibid.

21. Sarah Josepha Hale, *The Good Housekeeper* (Boston: Otis, Broaders, and Company, 1841; facsimile ed., Mineola, NY: Dover Publications, 1996), 88.

22. Hale, *The Good Housekeeper*, 83.

Chapter 6

1. Prosper Montaigne, ed., *Larousse Gastronomique* (New York: Clarkson Potter, 2001), 54.

2. Robert Chadwell Williams, *Horace Greeley: Champion of American Freedom* (New York: New York University Press, 2006), 281. See also the documents related to the purchase of Alaska at the Library of Congress at www.loc.gov/rr/program/bib/ourdocs/Alaska .html (accessed July 11, 2014).

3. Timothy J. Gilfoyle, *City of Eros: New York City, Prostitution, and the Commercialization of Sex, 1790–1820* (New York: W. W. Norton, 1994) 203; Obituary of Alexander Williams. "Williams, 'Ex-Czar' of Tenderloin, Dies," *New York Times*, March 26, 1917. www.newyorktimes.com (accessed July 10, 2014).

4. Cindy Lobel, "Sylvester Graham and Antebellum Diet Reform," Gilder-Lehrman Institute of American History, www.gilderlehrman.org/history-by-era/first-age-reform/essays /sylvester-graham-and-antebellum-diet-reform (accessed March 31, 2014).

5. Andrew F. Smith, *Eating History: Thirty Turning Points in the Making of American Cuisine* (New York: Columbia University Press, 2011), 127.

6. Steven Mintz and Susan Kellogg, *Domestic Revolutions: A Social History of American Family Life*, (New York: The Free Press, 1989), 83–87. See also Mercedes Sowko Crispin, "Carpatho-Rusyn Immigrants of Pennsylvania's Steel Mills, 1880–1920" (Humboldt State University Master Thesis, submitted May 2006), 13.

7. Mintz and Kellogg, *Domestic Revolutions*, 27.

8. Mintz and Kellogg, *Domestic Revolutions*, 29.

9. William Wallace Weaver, *Sauerkraut Yankees: Pennsylvania Dutch Foods and Foodways* (Mechanicsburg, PA: Stackpole Books, 2002), 42.

10. Elizabeth Nicolson, *The Economical Cook and House-Book* (Philadelphia: Ashmead and Evans, 1865), 18.

11. Mervyn Rothstein, "The Circle of Life with Bagels." *New York Times*, November 25, 2008. www.nytimes.com/2008/11/26/dining/26bagel.html?pagewanted=all.

12. Mimi Sheraton, *The Bialy Eaters: A Story of a Bread and a Lost World* (New York: Broadway Books, 2000), Loc 1183–1262. Also, Nina Selin, oral history interview conducted by the author, September 2013. Many thanks to Nina Selin, a descendant of Bialy bakers from Poland, for generously giving of her time for an interview. For more information about the richly complex and evolving history of Manhattan's Lower East Side, visit the Lower East Side Tenement Museum (www.tenement.org) in New York City.

13. The Well-Heeled Cook, "Celery," March 23, 2012, www.thewellheeledcook.com (accessed March 23, 2014). See also *Reading Eagle*, "Amish Wedding Recipes," November 24, 2010, www2.readingeagle.com/article.aspx?id=266942 and Amish Family Values, www.amishfamilyvalues.com.

14. Kenneth F. Kiple and Kriemhild Conee Ornelas, eds., "Celery and Celery Root," in *The Cambridge World History of Food*, Vol. 2 (Cambridge: Cambridge University Press, 2000), 1748–49.

15. Sylvia Lovegren, "Barbecue" in *The Oxford Encyclopedia of Food and Drink in America*, ed. Andrew F. Smith, Vol. 1 (New York: Oxford University Press, 2004), 64.

16. DeWitt, *The Founding Foodies*, 127 (see chap. 2, n.1).

17. The first printed recipe for Hoppin' John appeared in *The Carolina Housewife* by Sarah Rutledge in 1847, but the dish predates the publication of that book. Rice was an important crop for the coastal areas of South Carolina, and enslaved workers from Africa brought their native agricultural experience with rice growing to the New World during their forced migration, contributing to the immense prosperity of southern rice plantations. Black-eyed peas are indigenous to West Africa, original home of many of the people kidnapped by slave traders. I believe that West Africans ate black-eyed peas and rice (Hoppin' John) long before they came to the Carolinas. See Sarah Rutledge, *The Carolina Housewife* or *House and Home: by a Lady of Charleston*. First published 1847 (Columbia: University of South Carolina Press, 1979 reprint), 83. Today it is considered good luck to eat Hoppin' John on New Years Day. The origin of the name "Hoppin' John" may come from the French Creole name for black-eyed peas, *pois de pigeon*. Smith, *Food and Drink in America*, 1: 72.

CHAPTER 7

1. Upton Sinclair, "What Life Means to Me," *Cosmopolitan Magazine* 41 (October 1906): 594.

2. Smith, *Eating History*, 158.

3. Civitello, *Cuisine and Culture*, 303–4 (see chap. 1, n. 1).

4. Smith, *Eating History*, 100–102. Smith presents a wealth of material on "Fair Food." See also Daniel E. Sutherland, *The Expansion of Everyday Life, 1860-1876* (New York: Harper Row, 1989), 263–70. Also see Thomas J. Schlereth, *Victorian America: Transformations in Everyday Life* (New York: HarperCollins, 1991), 169–75. The study of America's world's fairs and expositions is a particularly rich area of historical research since these events presented America as it viewed itself through the lens of a civic booster. Demeaning exhibits featuring racial minorities were rife with prejudice and insults, but few white commentators even noticed. (See Schlereth, *Victorian America*, 173.)

5. Barry H. Landau, *The President's Table: Two Hundred Years of Dining and Diplomacy* (New York: HarperCollins, 2007), 185–86.

6. Sarah Tyson Rorer, *Mrs. Rorer's New Cook Book* (Philadelphia: Arnold and Company, 1902), 696–97. Sarah Rorer was a hugely popular cookbook "authoress" for a generation stretching from the late nineteenth century until the 1920s. See also Andrew F. Smith, *The Oxford Companion to American Food and Drink* (New York: Oxford University Press, 2007), 33–34.

7. Rich Cohen, *The Fish That Ate the Whale: The Life and Times of America's Banana King* (New York: Picador, 2013), 48–49.

8. There are many resources on Ellis Island, but for a reliable introduction, visit www.ellisisland.org/genealogy/ellis_island_timeline.asp (accessed December 3, 2013).

9. David M. Brownstone, Irene M. Franck, and Douglass Brownstone, *Island of Hope, Island of Tears: The Story of Those Who Entered the New World through Ellis Island* (New York: Barnes & Noble Books, 2000), 180–82.

10. David Wilma, "Automated Salmon Cleaning Machine," History Link, January 1, 2000, www.HistoryLink.org (accessed December 30, 2013).

11. Tannahill, *Food in History*, 310–13 (see chap. 2, n. 7).

12. Schlereth, *Victorian America*, 157.

13. Wikipedia's entry for Heinz 57 is excellent. See Wikipedia, "Heinz 57," en.wikipedia.org/wiki/Heinz_57. See also Mary Bellis, "Hmm Hmm Good: The Trademarks and History of Campbell's Soup," en.wikipedia.org/wiki/Heinz_57. Also see H. J. Heinz Company, www.heinz.com.

14. Hoover headed the Commission for the Relief of Belgium before the United States entered the war. Because the commission was so successful, Wilson appointed Hoover to the position of director of the U.S. Food Administration. See the Hoover Archives, "Gallery Two: The Humanitarian Years," hoover.archives.gov/exhibits/Hooverstory/gallery02/index.html.

15. Rae Katherine Eighmey, *Food Will Win the War: Minnesota Crops, Cooks, and Conservation during World War I* (St. Paul: Minnesota Historical Society Press, 2010), xi.

16. *Farmer's Wife*, May 1917, 270, cited in ibid., 23.

17. Diner, *Hungering for America*, Loc 82 of 3746, Loc 726 of 3746.

18. Amelia Doddridge, *Liberty Recipes* (New York: Stewart and Kidd, 1918; facsimile ed., Charleston, SC: Nabu Press, 2013), 49–50, 58–60.

19. *Good Housekeeping*, "Kitchen Soldier Pledge," March 1918, 51.

20. Claudia Cranston, "Uncle Sam Talks through Good Housekeeping," *Good Housekeeping*, September 1918, 54.

21. Army Heritage Center Foundation, "World War I Rations: Full Belly, Full Ready," armyheritage.org/education-and-programs/educational-resources/education-materials-index/50-information/soldier-stories/278-worldwar1rations (accessed October 22, 2013).

22. In total, 8.5 million people were killed in World War I, while 20 to 40 million were killed by the Spanish influenza epidemic, which still fascinates scientists. Carl Zimmer, "In 1918, Bad Timing Fed a Pandemic." *New York Times*, May 6, 2014. See American Red Cross, "A Brief History of the American Red Cross," www.americanredcross.org/about-us/history and PBS, "The Great War and the Shaping of the 20th Century," www.pbs.org/greatwar. (accessed March 25, 2014).

23. Eighmey, *Food Will Win the War*, 194.

24. Susan Strasser, *Never Done: A History of American Housework* (New York: Holt Paperbacks, 2000), x.

25. Emily Contois, "The Woman Suffrage Cookbook of 1886," Emilycontois.com.

26. "Alice Paul and the Women's Suffrage Movement," www.historywired.si.edu (accessed March 4, 2014). Eleanor Flexner classic *Century of Struggle* still presents a powerful overview and it's still in print. Eleanor Flexner, *Century of Struggle* (Cambridge, MA: Belknap Press, 1959). Visit the Sewall Belmont House in Washington, DC, home of the National Women's Party and now a museum that presents the story of women's suffrage in a clear, cohesive way.

27. Mary Beth Norton et al., *A People and a Nation: A History of the United States*, 4th ed. (Boston: Houghton Mifflin Company, 1994), 644, 703–705.

28. *The Balance and Columbian Repository*, "Communication to the Editor of the Balance," May 13, 1806, 146. www.imbibemagazine.com/images/Balance_5-13-1806.pdf as cited in Wikipedia, "Cocktail," en.wikipedia.org/wiki/Cocktail.

29. *The Balance and Columbian Repository*, 146.

30. Wikipedia, "Cocktail Party," en.wikipedia.org/wiki/Cocktail_party.

31. Norton et al., *A People and a Nation*, 740–44.

32. The size of the servings was smaller than they usually are today, resembling something like a lengthy tasting menu. This presentation of different foods for different courses was called "service à la Russe," or Russian-style service. Previously, in the older style, "service à la Francaise," many dishes were placed on the table at once. See Louise Belden, *The Festive Tradition: Table Decoration and Desserts in America, 1650–1900* (New York: W. W. Norton, 1983), 33–38.

33. Peter Smith, "Frito Pie and the Chip Technology that Changed the World," posted on *Food and Think*, *Smithsonian*, January 2012, www.blogs.smithsonianmag.com (accessed November 23, 2013). See also www.angelfire.com/tx2/martinez/chip.html (accessed March 25, 2014).

34. Gustavo Arellano, *Taco USA: How Mexican Food Conquered America* (New York: Simon and Schuster, 2012), 205–6.

35. Although by the mid-1700s, a Scottish naval surgeon, James Lind, understood that citrus fruit helped prevent the disease called scurvy, vitamin C wasn't identified until 1912. See Kevin J. Carpenter, "The Nobel Prize and the Discovery of Vitamins," Nobel Prize website, www.nobelprize.org/nobel_prizes/themes/medicine/carpenter/ (accessed March 26, 2014).

Chapter 8

1. Norton et al., *A People and a Nation*, 754.

2. Nicole Rittenmeyer, producer, *The Road to Civil Rights*, 2013, HISTORY. www.biography.com/people/zora-neale-hurston-9347659.

3. Harris, *High on the Hog*, 170–75 (see chap. 4, n. 13). See also Linda R. Monk and Ken Burns, *Ordinary Americans: U.S. History through the Eyes of Everyday People* (Alexandria, VA: Close Up Foundation, 1994), 178.

4. Zora Neale Hurston, *Their Eyes Were Watching God* (New York: HarperCollins e-books, 2013), 4. First published in 1937 by J. B. Lippincott.

5. Harris, *High on the Hog*, 175–76.

6. Harriet Ross Colquitt, *The Savannah Cook Book* (Charleston, SC: Colonial Publishers, 1933), 76.

7. Norton et al., *A People and a Nation*, 756–65.

8. Now called the Capuchin Service Center, this organization is still serving free meals. See Capuchin Soup Kitchen, www.cskdetroit.org.

9. Mark Kurlansky, ed., *The Food of a Younger Land* (New York: Riverhead Books, 2009), 1; Arthur Schlessinger, *The Coming of the New Deal, 1933–1935* (New York: Mariner Books, 2003), 180. Also, Norton et al., *A People and a Nation*, 792–94.

10. Another WPA recipe is posted on Spare a Dime, spare-a-dime.weebly.com/1/post/2013/04/wpa-soup-a-recipe-for-hard-times.html.

11. Monk and Burns, *Ordinary Americans*, 180–84. See generally, Timothy Egan, *The Worst Hard Time: The Untold Story of Those Who Survived the Great American Dust Bowl* (New York: Mariner Books, 2006). See also Susan Werbe, executive producer, *Black Blizzard*, documentary film, HISTORY.

12. Other U.S. farming regions were devastated by the fall in agricultural prices, and farm animals and produce were destroyed on a large scale to keep market values up. The destruction failed to raise prices and deprived truly starving people of cheap food. In 1933, over six million pigs were slaughtered. Many of the carcasses were ground up and dumped on wasteland or used as landfill in the Midwest. Janet Poppendieck, *Breadlines Knee Deep in Wheat: Food Assistance in the Great Depression* (New Brunswick, NJ: Rutgers University Press, 1986), 111–28.

13. Rita Van Amber, *Stories and Recipes of the Great Depression of the 1930s* (Neenah WI: Van Amber Publishers, 1986), 4 (Lamb's quarters), 5 (Dandelions), 150 (Milk soup).

14. Laura Shapiro, "The First Kitchen: Eleanor Roosevelt's Austerity Drive." *New Yorker*, November 22, 2010, www.newyorker.com/reporting/2010/11/22/101122fa_fact_shapiro.

15. Author's interview with Allida Black, founding editor of the *Papers of Eleanor Roosevelt*, and currently research professor of history and international studies at George Washington University. October 15, 2013. See also Landau, *The President's Table*, 182–83.

16. Wikipedia, "SPAM," Wikipedia.org/wiki/Spam_(food).

17. Ibid.

18. University of California, "Relocation and Incarceration of Japanese Americans during World War II," Japanese American Relocation Digital Archives (JARDA), www.calisphere.universityofcalifornia.edu/jarda/historical-context.html.

19. Nadia Arumgum, "From Budget Fare to Culinary Inspiration: The History of Meatloaf," *Atlantic Monthly*, September 20, 2011. www.theatlantic.com/health/archive/2011 (accessed November 14, 2013).

20. Ames (Iowa) Historical Society, "World War II Rationing on the U.S. Homefront," www.ameshistory.org/exhibits/events/rationing.htm.

21. "Rosie the Riveter: Women Working During World War II." www.nps.gov/pwro/collection/website/rosie.htm.

22. Civitello, *Cuisine and Culture*, 112, 138.

23. Wikipedia, "United States Military Chocolate," en.wikipedia.org/wiki/United_States_military_chocolate.

24. Smith, *Food and Drink in America*, 2: 245, 272.

25. Kevin Baker, *America The Story of Us* (New York: A&E Television Networks, 2010), 322–25; Lucas Films, producer, *V for Victory*, a documentary film, 2012.

26. Edna Lewis, *A Taste of Country Cooking* (New York: Alfred Knopf, 1976), 133–37.

27. Naval History and Heritage Command, "Navajo Code Talkers: World War II Fact Sheet," www.history.navy.mil/faqs/faq61-2.htm. See also Wikipedia, "Code Talker," www.en.wikipedia.org/wiki/Code_Talker. Also see History Channel, *Save Our History: Navaho Code Talkers* (1999) and U.S. Central Intelligence Agency, "Navajo Code Talkers and the Unbreakable Code, www.cia.gov/news-information/featured-story-archive/2008-featured-story-archive/navajo-code-talkers/ (accessed November 26, 2013).

28. U.S. Central Intelligence Agency, "Navajo Code Talkers and the Unbreakable Code, www.cia.gov/news-information/featured-story-archive/2008-featured-story-archive/navajo-code-talkers/ (accessed November 26, 2013).

29. Jen Miller, "Frybread," *Smithsonian*, July 2008, www.smithsonianmag.com/arts-culture/frybread-79191/ (accessed November 26, 2013).

30. There are many sources on Birdseye. See Library of Congress, "Everyday Mysteries: Who Invented Frozen Food," www.loc.gov (accessed November 20, 2013). See also Biography, "Clarence Birdseye," www.biography.com (accessed November 25, 2013). Also see Smith, *Eating History*, xx.

CHAPTER 9

1. Mary Bellis, "The History of Jello," About.com, inventors.about.com/library/inventors/bljello.htm (accessed December 14, 2013). See also "Jell-O: America's Most Famous Dessert," Duke University Libraries Digital Collections, library.duke.edu/digitalcollections/eaa_CK0029/ (accessed November 7, 2013).

2. Landau, *The President's Table*, 183.

3. Wikipedia, "Lettuce," en.wikipedia.org/wiki/Lettuce (accessed December 14, 2013). See also Thomas Turini et al., "Iceberg Lettuce Production in California," University of California Vegetable Research and Information Center, 2011, www.anrcatalog.ucdavis.edu/pdf/7215.pdf. (accessed December 14, 2013).

4. Rick Tejada-Flores, "Cesar Chavez & the UFW: The United Farmworkers Union," from *The Fight in the Fields*, PBS, 2004, www.pbs.org/itvs/fightfields/cesarchavez1.html. (accessed December 14, 2013).

5. The World of Coca-Cola, "Coca-Cola History," www.worldofcoca-cola.com/coca-cola-facts/coca-cola-history/ (accessed December 1, 2013).

6. Civitello, *Cuisine and Culture*, 242–44.

7. Coke cost five cents a serving for more than seventy years, partially because the original contract with the bottlers fixed the price. See Tim Harford, "The Mystery of the 5-cent Coca-Cola," *Slate*, May 11, 2007, www.slate.com/articles/arts/the_undercover_economist/2007/05/the_mystery_of_the_5cent_cocacola.html (accessed December 1, 2013).

8. Sally Edelstein, "On the Front Lines with Coca Cola, Part II," *Envisioning the American Dream*, May 30, 2013, envisioningtheamericandream.com/2013/05/30/on-the-front-lines-with-coca-cola-pt-ii/ (accessed December 1, 2013).

9. Elvis Presley was famous for his love of peanut butter and banana sandwiches, but this Jell-O salad cried out for recognition. "Salad: Sloppy, Soupy Concoction Wasn't Exactly Fit for the King," *Spokesman-Review*, November 1, 1994, news.google.com/newspapers?nid=1314&dat=19941101&id=SLQpAAAAIBAJ&sjid=PPEDAAAAIBAJ&pg=6839,564899 (accessed December 3, 2013). I'm so glad about the marshmallows in the recipe because, God knows, there isn't enough sugar

in that fabulous "salad" without them. I don't know if that recipe was Coke's idea, his grandmother's, or that of his manager, Colonel Parker. My money is on the colonel. For other Coke recipes, see for example, www.coca-colacompany.com/food/.

10. Hanna Miller, "American Pie," *American Heritage* 57, no. 2 (April/May 2006), xx, www.americanheritage.com/content/american-pie. See also Wikipedia, "Pizza," en.wikipedia.org/wiki/pizza (accessed December 14, 2013).

11. For more "fun facts" on pizza consumption, see www.statisticbrain.com/pizza.

12. Owen Edwards, "Tray Bon!" *Smithsonian*, December 2004, www.smithsonianmag.com /history-archaeology/Tray_Bon.html (accessed December 14, 2013).

13. "Postwar American Television: Estimated U.S. TV Sets and Stations," Early Television Museum, www.earlytelevision.org/us_tv_sets.html (accessed December 14, 2013).

14. In 2001, the Smithsonian National Museum of American History acquired a tin lunch box with an image of John Glenn and his space capsule on the front as part of its collection on popular culture. The lunch box was produced by the Thermos company in 1963. See National Museum of American History, "Orbit Lunch Box," americanhistory.si.edu /collections/search/object/nmah_1196973.

15. Miriam Kramer, "To Moonwalker Buzz Aldrin, 'Tang Sucks,'" SPACE.com, June 12, 2013, www.space.com/21538-buzz-aldrin-tang-spaceflight.html (accessed November 8, 2013).

16. Steve Brawley, "Jackie's Desserts," Pink Pill Box, www.pinkpillbox.com/desserts.htm (accessed December 14, 2013). See also Francois Rysavy, *A Treasury of White House Cooking* (New York: G. P. Putnam's Sons, 1972), 240. Also see Landau, *The President's Table*, 206–13.

17. David Halberstam, *The Fifties* (New York: Random House, 1993), 162, 164.

18. For more international McDonalds offerings, see www.buzzfeed.com/search?q=45-mcdonalds-items-not-available-in-the.+US+that+Should+be (accessed March 29, 2014).

19. Slow Food USA, www.slowfoodusa.org (accessed December 14, 2013). Full disclosure—I am a member of Slow Food USA, which is naturally headquartered in Brooklyn, New York. Berkeley, California, would have made sense too. The global headquarters are in Italy.

20. Smith, *Eating History*, 227.

21. "Interview with Georgia Gilmore" conducted by Blackside, Inc. on February 17, 1986, for *Eyes on the Prize: America's Civil Rights Years (1954–1965)*, Washington University Libraries, Film and Media Archive, Henry Hampton Collection, library.wustl.edu /units/spec/filmandmedia/collections/henry-hampton-collection/eyes1/gilmore.htm (accessed November 29, 2013).

22. Davia Nelson and Nikki Silva with Jay Allison, *Hidden Kitchens: Stories and More from NPR's The Kitchen Sisters*, narrated by Frances McDormand (Audio Renaissance, 2005), Disk 3, Track 6 of 14.

23. Ibid.

24. Joseph E. Holloway, "African Crops and Slave Cuisines," The Slave Rebellion website, slaverebellion.org/index.php?page=crops-slave-cuisines (accessed December 21, 2013). See also Thomas J. Craughwell, *Jefferson's Crème Brulee: How a Founding Father and His Slave James Hemings Introduced French Cuisine to America* (Philadelphia: Quirk Books, 2012), 189.

25. Interview with Martha Hawkins at "Martha's Kitchen," Biography, www.biography.com /people/martha-hawkins-21406965.

CHAPTER 10

1. Smith, *Food and Drink in America*, 1: 308.

2. Arthur Bartlett, "Popcorn Crazy," *Reader's Digest*, November 1949, 61–63.

3. D. R. McConnell, "The Impact of Microwaves on the Future of the Food Industry: Domestic and Commercial Microwave Ovens," *Journal of Microwave Power* 8, no. 2 (1973), www.jmpee.org/JMPEE_PDFs/08-2_bl/JMPEE-Vol8-Pg123-McConnell.pdf (accessed November 10, 2013).

4. Daniel J. DeNoon, "Microwave Popcorn Linked to Lung Harm," WebMD Health News, September 5, 2007, www.webmd.com/lung/news/20070905/microwave-popcorn-linked-to-lung-harm (accessed December 9, 2013).

5. Elizabeth David, *English Bread and Yeast Cookery* (New York: Penguin, 1980), 155–57.

6. Aaron Bobrow-Strain, "What Would Great-Grandma Eat," *Chronicle of Higher Education*, February 26, 2012, chronicle.com/article/What-Would-Great-Grandma-Eat-/130890/ (accessed November 9, 2013). See also Aaron Bobrow-Strain, *White Bread: A Social History of the Store-Bought Loaf* (Boston: Beacon Press, 2012).

7. Colin Schultz, "The Life and Death of Wonder Bread," *Smithsonian*, November 16, 2012, blogs.smithsonianmag.com/smartnews/2012/11/the-life-and-death-of-wonder-bread (accessed December 3, 2013).

8. Professor Jef I. Richards now teaches advertising at Michigan State University. cas.msu.edu/places/departments/advertising-pr/faculty-staff/name/jef-richards/.

9. Lisa Bramen, "Woodstock—How to Feed 400,000 Hungry Hippies," *Smithsonian*, August 14, 2009, blogs.smithsonianmag.com/food/2009/08/woodstock%E2%80%94how-to-feed-400000-hungry-hippies/ (accessed December 4, 2013).

10. David Kamp, *The United States of Arugula: How We Became a Gourmet Nation* (New York: Broadway Books, 2006). Electronic version, Loc 2511 of 6799.

11. Craughwell, *Thomas Jefferson's Crème Brûlée*, 181–82 (see chap. 9, n. 24). The sesame plants were originally from Africa.

12. Alice Waters, *Chez Panisse Menu Cookbook* (New York: Random House, 1982), 164. There is a lot of material online about Alice Waters, who is remarkably polarizing.

13. Marian Burros, "Eating Well; Old Lettuce Masquerades As Mesclun," *New York Times*, June 26, 1996, www.nytimes.com/1996/06/26/garden/eating-well-old-lettuce-masquerades-as-mesclun.html (accessed November 25, 2013).

14. amfAR, The Foundation for AIDS Research, "Thirty Years of HIV/AIDS: Snapshots of an Epidemic," www.Amfar.org, www.amfar.org/thirty-years-of-hiv/aids-snapshots-of-an-epidemic/ (accessed November 22, 2013).

15. Centers for Disease Control and Prevention, "HIV in the United States: At a Glance," www.cdc.gov/hiv/statistics/basics/ataglance.html (accessed November 24, 2013).

16. "First Wave Feminism" was women's fight for the vote. For more on the 1970 protest, see David M. Dismore, "When Women Went on Strike: Remembering Equality Day, 1970," *Ms. Magazine Blog*, August 10, 2010, msmagazine.com/blog/2010/08/26/when-women-went-on-strike-remembering-equality-day-1970/ (accessed December 9, 2013). Second-wave feminism still faces obstacles related to their goals of quality day care and equal pay. See the Lily Ledbetter Fair Pay Act, signed by President Obama in 2009, based on Title VII of the 1964 Civil Rights Act.

17. George M. Taber, "Modern Living: Judgment of Paris," *Time Magazine*, June 7, 1976, content.time.com/time/magazine/article/0,9171,947719,00.html.

18. Patrick E. McGovern, in Dolores R. Pipermo, ed., "Beginning of Viniculture in France," Smithsonian National Museum of Natural History and Smithsonian Tropical Research Institute (Fairfax, Washington, DC.) Approved May 1, 2013, www.pnas .org/content/early/2013/05/30/1216126110.abstract (accessed May 1, 2013).

19. Craughwell, *Thomas Jefferson's Crème Brûlée*, 12.

20. Adam Cole and Helen Thompson, "Archaeologists Find Ancient Evidence of Cheese-making," NPR The Salt: What's on Your Plate? December 13, 2012, www.npr.org /blogs/thesalt/2012/12/13/167034734/archaeologists-find-ancient-evidence-of-cheese-making (accessed December 13, 2013).

21. University of California, Davis, "The True Story of Monterey Jack Cheese," Dairy Research and Information Center, drinc.ucdavis.edu/dfoods4_new.htm (accessed December 12, 2013).

22. See specifically, the film segment produced by Kraft in 1959 on display at the Smithsonian's National Museum of American History's exhibition, "Food: From Field to Table."

23. Ronald Reagan, quoted in a Farm Service Agency news release, U.S. Department of Agriculture Fram Service Agency, "Commodity Credit Corporation," November 1999. www.fsa.usda.gov/FSA/newsReleases?area=newsroom&subject=landing&topic =pfs&newstype=prfactsheet&type=detail&item=pf_19991101_comop_en_ccc.html (accessed December 20, 2013).

24. Janet Poppendieck, *Free for All: Fixing School Food in America* (Berkeley: University of California Press, 2010), 76.

25. Junot Diaz, "How to Date a Brown Girl (Black Girl, White Girl, or Halfie)." *New Yorker*, December 25, 1995, 83.

26. Topix: Jamestown Forum, "Remember Commodity Cheese?" www.topix.com/forum/city /jamestown-tn/TIGPNSSUQDTVUUM58 (accessed December 11, 2013).

27. Parke Wilde, comment on "Despite Dietary Guidelines, Federal Government Promotes High-Fat Cheese through Dairy Checkoff," U.S. Food Policy Blogspot, November 7, 2010, usfoodpolicy.blogspot.com/2010/11/despite-dietary-guidelines-federal.html (accessed December 13, 2013).

28. Michael Moss, "The Day They Took the Cheese out of Cheez Whiz," *National Post*, March 9, 2013, fullcomment.nationalpost.com/2013/03/09/michael-moss-the-day-they-took-the-cheese-out-of-cheez-whiz/ (accessed December 13, 2013).

29. Molly O'Neill, "New Mainstream: Hot Dogs, Apple Pie and Salsa," *New York Times*, March 11, 1992, www.nytimes.com/1992/03/11/garden/new-mainstream-hot-dogs-apple-pie-and-salsa.html?pagewanted=all&src=pm (accessed December 13, 2013).

30. Ibid.

31. Carl Bialik, "Ketchup vs. Salsa: By the Numbers," *Wall Street Journal*, "The Numbers Guy," (September 20, 2007), blogs.wsj.com/numbersguy/ketchup-vs-salsa-by-the-numbers-191/ (accessed December 4, 2013).

32. Ibid.

33. Jessica Chen, "Biting into Sushi," *Entreprenuer*, December 11, 2007, www.entrepreneur .com/article/188012 (accessed December 7, 2013).

34. *Food & Wine*, "Sushi in America," September 2005, www.foodandwine.com/articles /sushi-in-america (accessed December 7, 2013).

35. Alastair Bland, "From Cat Food to Sushi Counter: The Strange Rise of Bluefin Tuna," *Smithsonian*, September 11, 2013, blogs.smithsonianmag.com/food/2013/09/from-

cat-food-to-sushi-counter-the-strange-rise-of-the-bluefin-tuna/ (accessed December 7, 2013). Bland also discusses how the popularity of sushi has sadly damaged the global fish supply, particularly bluefin tuna.

EPILOGUE

1. Linda Stradley, "History of Chili, Chili Con Carne," whatscookingamerica.net/History /Chili/ChiliHistory (accessed February 4, 2014). For the history of the Engine Company 33 and Ladder Company 9 firehouse, see Wikipedia, en.wikipedia.org /wiki/Firehouse,_Engine_Company_33_and_Ladder_Company_9.

2. Derivation of "jonesing" from English Language and Usage Stack Exchange, www.english .stackexchange.com.

3. Oprah.com, "Dr. Perricone's 10 Superfoods You Should Add to Your Diet Today," July 15, 2005, www.Oprah.com (accessed February 13, 2014). See also *Dr. Oz Show*, "Superfood to Boost Your Immunity: Seaweed," video segment, October 21, 2013, www.doctoroz.com/slideshow/dr-ozs-10-favorite-superfoods?gallery=true&page=11 (accessed February 13, 2014).

4. BBC News, "SuperFoods 'Ban' Goes into Effect," June 29, 2007, news.bbc.co.uk/2/hi /health/6252390.stm (accessed February 13, 2014). See also Amelia Hill, "Forget Superfoods: You Can't Beat an Apple a Day," *Guardian*, May 12, 2007, www.theguardian .com/uk/2007/may/13/health.healthandwellbeing1 (accessed February 13, 2014).

5. Lesley Rotchford, "Diets through History: The Good, the Bad, and the Scary," CNN Health, February 8, 2013, www.cnn.com/2013/02/08/health/diets-through-history/ (accessed February 13, 2004). See also Mirielle Guiliano, *French Women Don't Get Fat* (New York: Knopf, 2004).

6. Simple temperature changes the molecular structure of food, so the fact that people are changing food molecules is nothing new. For example, raw egg molecules change in heat, turning transparent egg whites opaque. Crystals from frozen milk represent molecular changes caused by freezing. Molecular gastronomy applies naturally occurring chemicals to change the molecular structure. In contrast, GMO foods have their DNA changed in ways that would have been impossible naturally, which permanently alters the organism because that change is passed on genetically.

7. There are kits you can buy to replicate these results which are instructional and a lot of fun. See www.Molecule-R.com.

8. Greg Morabito, "15 New York Chefs Who Don't Like 'Molecular Gastronomy'" March 7, 2011 (search "molecular gastronomy" on www.ny.eater.com). The quote from Anthony Bourdain was accessed March 12, 2014.

9. Linda Stradley, "Apples, History and Legends of Apples," 2004, whatscookingamerica.net /Fruit/Apples.htm (accessed March 15, 2014).

10. To put this in perspective, the number of people in developing countries who die every year from vitamin A deficiencies is only 10 percent less than the 1.7 million who die from HIV-AIDS annually. See Wikipedia, "Golden Rice," en.wikipedia.org/wiki/Golden_rice.

11. UNICEF, "Vitamin A Deficiency," www.childinfo.org/vitamina.html. See also David Ropeik, "Golden Rice Opponents Should Be Held Accountable for Health Problems Linked to Vitamin A Deficiency," *Scientific American*, March 15, 2014, blogs.scientificamerican. com/guest-blog/2014/03/15/golden-rice-opponents-should-be-held-accountable-for-health-problems-linked-to-vitamain-a-deficiency/ (accessed March 17, 2014).

PHOTO CREDITS

CHAPTER 7

CHAPTER 8

CHAPTER 9

CHAPTER 10

EPILOGUE

INDEX

❧ INDEX OF RECIPIES ❧

✺ ABOUT THE AUTHOR ✺

Libby H. O'Connell is chief historian for History Channel and senior vice president for corporate social responsibility for A+E Networks, where she has worked since 1993. She develops educational and community-based initiatives for A+E Networks. As a historical adviser for the History programming group, she has appeared as a commentator on History and A&E, as well as on the *Today Show*, CNN, and other news channels. She pro-

Photo by Tess Steinkolk

duces on-site short films for organizations such as the Smithsonian, the National Park Service, National Archives, and the Library of Congress. Her video blogs on History.com focus on historical stories of place and food. Currently, she serves on the boards of Thomas Jefferson's Monticello, National History Day, and the Smithsonian National Museum of American History's "Kitchen Cabinet," a group dedicated to the study and sharing of food history. Dr. O'Connell's work in television, historic preservation, and education has received three national Emmy awards and numerous other honors. She received her Ph.D. in American history from the University of Virginia. She lives outside of New York City.